Raphael Tautz

Charge Separation in Organic Photovoltaics

Raphael Tautz

Charge Separation in Organic Photovoltaics

Enhanced Formation of Weakly Bound Polaron Pairs in Donor-Acceptor-Copolymers

Südwestdeutscher Verlag für Hochschulschriften

Impressum / Imprint

Bibliografische Information der Deutschen Nationalbibliothek: Die Deutsche Nationalbibliothek verzeichnet diese Publikation in der Deutschen Nationalbibliografie; detaillierte bibliografische Daten sind im Internet über http://dnb.d-nb.de abrufbar.
Alle in diesem Buch genannten Marken und Produktnamen unterliegen warenzeichen-, marken- oder patentrechtlichem Schutz bzw. sind Warenzeichen oder eingetragene Warenzeichen der jeweiligen Inhaber. Die Wiedergabe von Marken, Produktnamen, Gebrauchsnamen, Handelsnamen, Warenbezeichnungen u.s.w. in diesem Werk berechtigt auch ohne besondere Kennzeichnung nicht zu der Annahme, dass solche Namen im Sinne der Warenzeichen- und Markenschutzgesetzgebung als frei zu betrachten wären und daher von jedermann benutzt werden dürften.

Bibliographic information published by the Deutsche Nationalbibliothek: The Deutsche Nationalbibliothek lists this publication in the Deutsche Nationalbibliografie; detailed bibliographic data are available in the Internet at http://dnb.d-nb.de.
Any brand names and product names mentioned in this book are subject to trademark, brand or patent protection and are trademarks or registered trademarks of their respective holders. The use of brand names, product names, common names, trade names, product descriptions etc. even without a particular marking in this works is in no way to be construed to mean that such names may be regarded as unrestricted in respect of trademark and brand protection legislation and could thus be used by anyone.

Coverbild / Cover image: www.ingimage.com

Verlag / Publisher:
Südwestdeutscher Verlag für Hochschulschriften
ist ein Imprint der / is a trademark of
AV Akademikerverlag GmbH & Co. KG
Heinrich-Böcking-Str. 6-8, 66121 Saarbrücken, Deutschland / Germany
Email: info@svh-verlag.de

Herstellung: siehe letzte Seite /
Printed at: see last page
ISBN: 978-3-8381-3683-7

Zugl. / Approved by: München, LMU München, Dissertation, 2012

Kurzfassung

Polymere mit Halbleiter-Eigenschaften haben ein großes Anwendungspotential in der organischen Photovoltaik, da sich ihre optischen und elektronischen Eigenschaften über die molekulare Struktur gezielt ändern lassen. Durch die Synthese von Copolymeren mit besonders kleiner optischer Bandlücke (low-bandgap Copolymere) konnte die Absorption von Sonnenlicht weiter in den infraroten Spektralbereich ausgedehnt und somit die Konversion von Sonnenlicht in elektrische Energie deutlich verbessert werden. Diese neuartigen Donor-Akzeptor Materialien basieren auf einer alternierenden Anordnung von elektronen-reichen und -armen Blöcken, die durch elektronische Kopplung neue Energieniveaus mit kleinerer optischer Bandlücke bilden.

Ziel dieser Arbeit ist die eingehende Untersuchung der photophysikalischen Eigenschaften dieser weitgehend unerforschten Moleküle. Die ersten drei Kapitel bieten dem Leser eine Einführung in das Forschungsgebiet und in die theoretische Beschreibung konjugierter Polymere, sowie einen Überblick über den aktuellen technischen Stand organischer Photovoltaik. Kapitel 4 gibt eine Zusammenfassung der verwendeten experimentellen und theoretischen Methoden.

Der erste Teil der Untersuchung von Donor-Akzeptor Materialien gilt den Photoanregungen und der korrekten Zuordnung ihrer spektralen Signaturen (Kap. 5). Diese ermöglicht eine Zuordnung der spektralen Signaturen zu stark gebundenen, elektrisch neutralen Exzitonen, bzw. leichter zu trennenden Ladungsträgerpaaren mit kleinerer Bindungsenergie, sogenannten Polaronenpaaren. Aufgrund der schwachen elektrischen Abschirmung von Ladungen in organischen Materialen liegen die meisten Photoanregungen als Exzitonen vor. In dieser Hinsicht zeigen spektroskopische Messungen auf Femtosekunden-Zeitskala erstmals den andersartigen Charakter von Donor-Akzeptor Materialien und demonstrieren den großen Einfluss ihrer Struktur auf die Art der erzeugten Photoanregungen. Sie zeigen, dass bei Photoanregungen dieser neuartigen Materialien neben Exzitonen auch ein beträchtlicher Anteil an Polaronenpaaren entsteht. Diese Donor-Akzeptor Materialien weisen einen Polaronenpaar-Anteil von bis zu 24% aller Photoanregungen auf, was dem Dreifachen der Effizienz vergleichbarer Homopolymere entspricht (Kap. 6). Weitere Untersuchungen zeigen außerdem eine erhöhte Erzeugungsrate bei kürzeren Anregungswellenlängen. Dies kann auf eine Korrelation mit einem ausgeprägten Elektronentransfer der involvierten Wellenfunktion zurückgeführt werden, welcher in theoretischen Simulationen deutlich wird (Kap. 7).

Zusammenfassend geben die in dieser Arbeit dargestellten Ergebnisse einen detaillierten Einblick in die optischen und elektronischen Eigenschaften von Donor-Akzeptor Copolymeren und den starken Einfluss der molekularen Struktur auf die ersten Schritte der photovoltaischen Stromerzeugung. Zusammenhänge zweier Schlüsselfaktoren für die Effizienzsteigerung zukünftiger organischer Solarzellen mit Materialparametern werden deutlich. Dies sind die Erzeugungseffizienz und die Lebensdauer von Polaronenpaaren und deren Abhängigkeit von der Elektronegativität und der Abstand von Akzeptor- zu benachbarten Donorsegmenten. Weiterhin konnte eine ausgeprägte Polaronenpaar Erzeugung über das ganze Absorptionsspektrum nachgewiesen werden. Diese Erkenntnisse bieten eine große Hilfestellung bei der weiteren Optimierung von Polymeren für Photovoltaik. Außerdem heben sie den wichtigen Beitrag der Ultrakurzzeit Spektroskopie zum grundlegenden Verständnis der Polaronenpaarerzeugung hervor. Mit diesen Mitteln könnte eine Verringerung des Spannungsverlustes möglich werden, der zur Ladungsträgertrennung in organischen Materialien nötig ist.

Abstract

Polymeric semiconductors attract a great deal of attention due to their ability to achieve unique optical and electronic properties by tailoring their chemical structure. This allows synthesizing fully conjugated, novel materials with suitable functionalities for organic electronics. For organic photovoltaics, the development of a novel class of materials significantly improved the infrared light absorption and resulted in a better coverage of the solar spectrum. These novel low-bandgap materials, i.e. donor-acceptor copolymers, are based on an alternating arrangement of electron donating and accepting moieties within their repeating unit. They exhibit an exceptionally enhanced absorption behavior what is attributed to their highly electronic coupling.

This work is focused on the physical properties of these novel and largely unexplored materials, with an emphasis on light absorption and charge carrier formation. In the first three chapters a proper introduction to the scientific field, the theoretical background and a brief overview on the state of the art of organic photovoltaics is given. Afterwards, the large variety of employed experimental and theoretical methods will be described in detail in the fourth chapter.

A careful study of the photoexcited states and their spectral signatures in donor-acceptor copolymers is the first step of their characterization, as presented in the fifth chapter. This allows the assignment of different spectral features to strongly bound, electrically neutral excitons, or to spatially separated charge carriers with weaker Coulomb interaction, called polaron pairs. Due to the weak dielectric screening, which is typical of most organic materials, strongly bound excitons are the major species of formed photoexcitations. However, polaron pairs promise easier dissociation and thus are of great interest regarding their contribution to photocurrent. Observing photoexcitations on femtosecond timescale by ultrafast spectroscopy reveals for the first time a strong impact of the alternating donor-acceptor units on the evolution of generated photoexcitations, as presented in the sixth chapter. In these novel copolymers, not only excitons are formed by light absorption, but also a substantial amount of weakly bound polaron pairs. The found polaron pair yields in donor-acceptor materials reach up to 24% and are thus about three times higher than for commonly used homopolymers. Further experiments, applying different excitation energies, discovered even increased polaron pair yields when exciting with significant excess energy. A strong correlation between polaron pair yield and a calculated charge transfer character of the involved electron wave functions is found to be the origin of this improved behavior, as detailed in the seventh chapter. Further localization due to spatial restrictions in small donor-acceptor molecules significantly increases this effect and are shown to exhibit even higher polaron pair yields.

Altogether, this study offers a detailed view on the optical and electronic properties of donor-acceptor copolymers. The results in this study illustrate the important role of chemical structure on the first steps of photovoltaic action, i.e. the formation of charge carriers and their dissociation. We can then relate two key factors for the success of future photovoltaic devices, namely polaron pair formation and their lifetime, to the material's chemical structure, i.e. the acceptor electron affinity, its separation to the neighboring donor units and the molecular chain length. Furthermore, a substantial polaron pair yield through the whole absorption spectrum of such materials is demonstrated. These results provide the first platform for rational designing of new materials for organic photovoltaics and place ultrafast spectroscopy as an important tool for understanding the governing factors of polaron pair formation, which is of utmost importance in this context. By that, a successful reduction of the required voltage loss for dissociation and extraction of charge carriers from organic materials might thus become reality in near future.

List of Publications

This thesis is based on the following publications:

Charge Photogeneration in Donor-Acceptor Conjugated Materials: Influence of Excess Excitation Energy and Chain Length

R. Tautz, E. Da Como, Ch. Wiebeler, G. Soavi, I. Dumsch, N. Fröhlich, G. Grancini, S. Allard, U. Scherf, G. Cerullo, S. Schumacher, J. Feldmann

The Journal of the American Chemical Society, *in press* (2013) –
DOI: 10.1021/ja309252a

Structural correlations in the generation of polaron pairs in low-bandgap polymers for photovoltaics

R. Tautz, E. Da Como, T. Limmer, J. Feldmann, H.-J. Egelhaaf, E. von Hauff, V. Lemaur, D. Beljonne, S. Yilmaz, I. Dumsch, S. Allard, U. Scherf

Nature Communications **3**, 970 (2012)

Spectral Signatures of Polarons in Conjugated Co-Polymers

Ch. Wiebeler, R. Tautz, J. Feldmann, E. von Hauff, E. Da Como, S. Schumacher

Journal of Physical Chemistry B, *in press* (2012) - DOI: 10.1021/jp3084869

Further scientific publications:

Influence of Carrier Density on the Electronic Cooling Channels of Bilayer Graphene

T. Limmer, A. J. Houtepen, A. Niggebaum, R. Tautz, and E. Da Como

Applied Physics Letters **99**, 3 (2011)

Reduced Charge Transfer Exciton Recombination in Organic Semiconductor Heterojunctions by Molecular Doping

F. Deschler, E. Da Como, T. Limmer, R. Tautz, T. Godde, M. Bayer, E. von Hauff, S. Yilmaz, S. Allard, U. Scherf, and J. Feldmann

Physical Review Letters **107**, 4 (2011)

Dispersion Control with Reflection Grisms of an Ultra-Broadband Spectrum Approaching a Full Octave

T. H. Dou, <u>R. Tautz</u>, X. Gu, G. Marcus, T. Feurer, F. Krausz, and L. Veisz

Optics Express **18**, 27900 (2011)

Toward Single Attosecond Pulses Using Harmonic Emission from Solid-Density Plasmas

P. Heissler, R. Horlein, M. Stafe, J. M. Mikhailova, Y. Nomura, D. Herrmann, <u>R. Tautz</u>, S. G. Rykovanov, I. B. Foldes, K. Varju, F. Tavella, A. Marcinkevicius, F. Krausz, L. Veisz, and G. D. Tsakiris

Applied Physics B-Lasers and Optics **101**, 511 (2010)

Approaching the Full Octave: Noncollinear Optical Parametric Chirped Pulse Amplification with Two-Color Pumping

D. Herrmann, C. Homann, <u>R. Tautz</u>, M. Scharrer, P. S. Russell, F. Krausz, L. Veisz, and E. Riedle

Optics Express **18**, 18752 (2010)

Investigation of Two-Beam-Pumped Noncollinear Optical Parametric Chirped-Pulse Amplification for the Generation of Few-Cycle Light Pulses

D. Herrmann, <u>R. Tautz</u>, F. Tavella, F. Krausz, and L. Veisz

Optics Express **18**, 4170-4183 (2010)

Density-Transition Based Electron Injector For Laser Driven Wakefield accelerators

K. Schmid, A. Buck, C. Sears, J. Mikhailova, <u>R. Tautz</u>, D. Herrmann, M. Geissler, F. Krausz, L. Veisz

Physical Review Special Topics - Accelerators and Beams **13**, 091301 (2010)

Generation of Sub-Three-Cycle, 16 TW Light Pulses by Using Noncollinear Optical Parametric Chirped-Pulse Amplification

D. Herrmann, L. Veisz, <u>R. Tautz</u>, F. Tavella, K. Schmid, V. Pervak, and F. Krausz, Optics Letters **34**, 2459 (2009)

Few-Cycle Laser-Driven Electron Acceleration

K. Schmid, L. Veisz, F. Tavella, S. Benavides, <u>R. Tautz</u>, D. Herrmann, A. Buck, B. Hidding, A. Marcinkevicius, U. Schramm, M. Geissler, J. Meyer-Ter-Vehn, D. Habs, and F. Krausz Physical Review Letters **102**, 4 (2009)

Laser-Driven Electron Acceleration in Plasmas with Few-Cycle Pulses

L. Veisz, K. Schmid, F. Tavella, S. Benavides, <u>R. Tautz</u>, D. Herrmann, A. Buck, B. Hidding, A. Marcinkevicius, U. Schramm, M. Geissler, J. Meyer-Ter-Vehn, D. Habs, and F. Krausz

Comptes Rendus Physique **10**, 140 (2009)

Contents

Contents

Contents

1. Introduction

"Die Sonne schickt uns keine Rechnung"
  Franz Alt

"The sun doesn't send us a bill". With this citation, the book author and moderator Franz Alt advertises in a very elegant way the free harvesting of solar energy [1]. Also because of contributions like this, the awareness has been growing in the past few decades, that a satisfaction of our ever increasing energy demands the exploration of new routes towards renewable sources. Typically, this insight is guided by obvious changes in our environment and the consequences for our every life, which are in turn the direct outcome of our extensive burning of non-renewable energy sources. These are increasing energy costs because of the shrinking oil reservoir, and also changes in our planet's climate due to vast emission of CO_2, as well as the radioactive contamination of large landscapes by nuclear breakdowns. To overcome the need of fossil or hazardous resources, large efforts have been made to increase the contribution of renewable sources to our supply of electrical energy. In the year 2011, the total renewable share of global energy consumption was 16.7% [2], including techniques like biomass, hydro, geothermal and solar energy harvesting. In Germany, photovoltaic devices exhibit as a part of these a pleasing growth rate, which allowed to increase its contribution from 3% in in the year 2011 to an expected value above 4.5% at the end of 2012 [3]. Nevertheless, to maintain its high growth rate and achieve a competitive price for solar energy, significant efforts in research and development are made toward an increase of efficiency and a reduction of costs for photovoltaic devices.

1

The latter goal has been mainly adressed by the development of 3rd generation solar cells, which include especially thin-film and organic solar cells. Although their efficiency is reduced compared to traditional silicon devices, thin-film architecture is less expensive and eases the requirements on material crystallinity and charge carrier mobility due to the shorter extraction length for charge carriers. Not only silicon based thin-film solar cells have been realized, but also successful devices based on organic materials, i.e. in special organic conjugated polymers and small molecules. These molecular materials combine semiconductivity and desirable optical properties with attractive physical characteristics, known from the field of plastics. The latter are mechanical flexibility, non-toxicity and the promising ability for large-scale processibility via roll-to-roll printing from solution. Moreover, these beneficial properties and the high potential for upscaling the fabrication size and speed make organic solar cells promising candidates for mass production [4]. Despite their promising properties, organic solar cells suffer from lower power conversion efficiencies than inorganic cells [5]. The record in terms of power conversion efficiency for these solar cells has recently been reported with 10% [5]. One of the main obstacles of organic solar cells to show better performance is the small dielectric constant of these materials, which is typically on the order of only 3.5 compared to 11 of silicon. This results in poor electric screening and thus a strong Coulomb interaction of charge carriers in such systems [6]. While this strong binding between charge carriers is regarded as beneficial for luminescent devices like organic light emitting diodes (OLEDs) [7], it is a crucial disadvantage for the aimed separation of photogenerated charge carrier pairs in organic solar cells [8]. Because of the large binding energy of photogenerated electron-hole pairs (> 0.5 eV), organic solar cells are typically based on bulk heterojunctions using a mixed blend of a suitable polymer and a derivative of the strongly electron-accepting Buckminster fullerene. A pronounced energy level offset provides the required driving force for charge separation at the interface between polymer and fullerene, which in turn results in a significant loss of usable solar cell voltage [9]. Light absorption and formation of strongly bound electron-hole pairs, so-called Frenkel excitons, is known to be almost exclusively performed by the polymer in the blend, while their dissociation still needs to be understood and has been subject of intensive recent investigations [6,10–14]. Several experimental and theoretical studies have investigated these first steps of charge carrier generation. Some of them suggested that among the primary photoexcitations are not only strongly bound Frenkel excitons, but also to a certain ratio spatially separated but still Coulomb bound charge carrier pairs, namely polaron pairs [15–17]. There are strong indications that the latter exhibit a weaker binding energy and thus require less driving force for their dissociation. Thus generating more polaron pairs and reducing their binding energy would be the first step in overcoming the detrimental strong Coulomb-coupling.

With this background, the focus of this work lies on the investigation of photophysical properties of novel materials for organic photovoltaics, i.e. low-bandgap copolymers [18,19]. By using these materials, highly efficient organic solar cells with record breaking power conversion efficiencies have been realized recently. These coplymers share a unique chemical structure, which is based on the alternating arrangement of electron donating and accepting moieties along the copolymer backbone. Electronic coupling between the two different moieties in donor-acceptor materials forms new energy levels, which result in a lower optical bandgap and consequently an improved absorption at long wavelengths. But this benefit alone cannot explain the exceptionally high efficiencies that are obtained with their usage. It clearly indicates that the understanding of the optical and electronic properties of these novel donor-acceptor materials is still very limited, and theoretical descriptions of common homopolymers fail in some parts when applied to these systems. Because of the high interest in revealing the basic processes governing charge carrier generation, in terms of physical knowledge and technical application, a thorough investigation with various experimental and theoretical methods has been employed. As first, the presented work will give a brief summary of the basic theoretical description of π-conjugated materials and an overview of the state of the art in organic photovoltaics. After giving detailed information about the experimental and theoretical methods, the following chapters will focus on spectroscopic signatures of excited states in organic materials, what is the basis for the further presented investigations. The main insights in the physical phenomena regarding donor-acceptor materials is presented in chapter six, where an enhanced formation of polaron pairs in donor-acceptor copolymers is attributed to the unique properties of their chemical structure. A detailed study about the role of excess excitation energy will complete this study and will allow to draw conclusion about the physical origin of polaron pair formation and to guide the way towards possible paths for its optimization. the reader is referred to the eighth chapter for the conclusions of this work and an outlook towards remaining open questions to be addressed in the future.

2. Theoretical description of π-conjugated materials

This chapter presents an introduction to the mathematical description of π-conjugated materials. It discusses the electronic structure of these materials and present different methods of their quantum mechanical description. Special emphasize lies on the detailed discussion of their optical properties and and the nature of photoexcitations in amorphous organic films. A thorough description of excitons, polarons and polaron pairs will familiarize the reader with these different kinds of photoexcitations and shed light on their dissociation and the formation of free charge carriers.

2.1. Electronic structure of π-conjugated materials

Organic semiconductor materials, as they are studied in this work, owe their electronic properties mainly to the molecular structure, than to bulk material properties, as it is in the case of inorganic semiconductors, such as silicon or germanium. Due to the large number of possible molecular bonds that carbon atoms can share, there exists a large variety of organic carbon-based materials with different electronic properties. This variety originates in the ability of carbon and several other chemical elements to form hybrid orbitals, allowing the formation of different molecular bonding types. In the ground state configuration of a carbon atom, the electrons occupy the shells in the configuration $1s^2\ 2s^2\ 2p^2$. When interacting with the electron orbitals of another atom, new favored energetic states are formed, these are known as hybrid

5

electron orbitals [20,21]. Given by the number of required covalent bonds, different grades of hybridization exist, which are well distinct in their geometric shape and the electronic properties of the resulting molecule. Since only the outer shell electrons are interacting in molecular bonds, only the $2s^2$ and the $2p^2$ electrons are available for hybrid orbitals. In the case of sp, sp^2 and sp^3 hybridization, two, three and four hybrid orbitals at a lower energetic position are formed, respectively, while the remaining non-hybridized outer shell electrons are present in regular $2p$ orbitals. These three cases lead to very distinct molecular geometries. A schematic overview of the formed occupied hybrid orbitals and their resulting orbital geometries can be found in Figure 2.1.

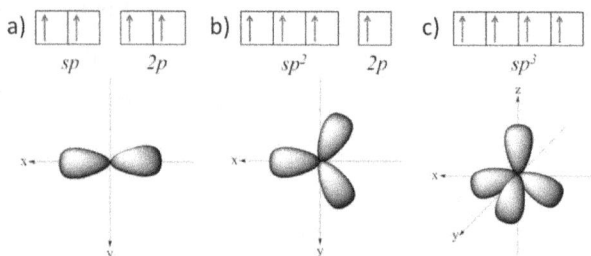

Figure 2.1: *Schematic overview of occupation and geometry of hybrid carbon orbitals. a) sp-hybridization results in a linear arrangement of two hybrid orbitals and two remaining occupied 2p orbitals. b) Three orbitals are formed at sp^2 hybridization. They show trigonal alignment with an angle of 120° between them. The remaining 2p orbital is arranged perpendicular to the hybrid orbital plane. c) sp^3 hybridization leads to a tetragonal orbital shape with no remaining 2p orbital. Visualizations taken from reference [22].*

The bonding of other atoms to carbon hybrid orbitals happens exclusively σ-bonds. Due to the electron wave function, which is centered along the connection axis between the nuclei of the contributing atoms, the overlap of the involved orbitals is very pronounced. This leads to an energetically favorable state. Therefore, the lowest electronic transitions between binding σ- and anti-binding σ^*-orbitals in those systems require a large amount of excitation energy, which is on the order of 3 eV. Electronic transitions in molecules based on sp^3 hybridized carbon thus require ultraviolet photons and do not typically provide interesting optoelectronic behavior upon irradiation with visible light. A well known representative of this class is ethane. As shown in Figure 2.3 c), both carbon atoms of ethane show a tetrahedral geometry, which is given by the typical alignment of the four equivalent sp^3 hybrid orbitals.

Carbon shows another different molecular geometry when sp^2 hybridized. This leads to a trigonal geometry, with a coplanar alignment of all three hybrid orbitals at an angle of 120° between them. Again, other atoms can form a σ-bonding with each of these orbitals. In this configuration, one remaining occupied $2p$ orbital is present, which is perpendicular to the plane of the hybrid orbitals. It allows other atoms with an occupied p-orbital to form an additional π-bond, whose electron wave function is off-centered along the connection line of the two nuclei. The reduced wave function overlap of the two p-orbitals results in a weaker bond than is usually known for σ-bonds. These double bonds, consisting of a σ- and a π-bond hence show electronic transitions between binding π and anti-binding π^*-orbitals ($\pi - \pi^*$) at moderate energies, which are accessible already at about 1.5 eV [23,24] (compare Figure 2.2).

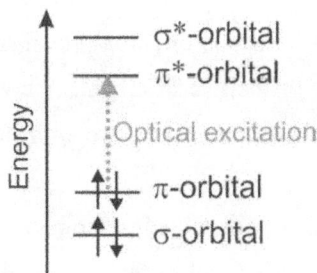

Figure 2.2: *Schematic energy diagram of molecular orbitals. In ground state, the binding σ- and π-orbitals are occupied while all anti-binding orbitals*

(marked with an asterisk) remain unoccupied. Electronic transitions between π and π^*orbitals are already accessible by optical excitation with photon energies as low as ca. 1.5 eV

An example for this system is the planar molecule ethylene, consisting of two planar CH_2 complexes connected by a $C=C$ double bond, as shown in Figure 2.3 b).

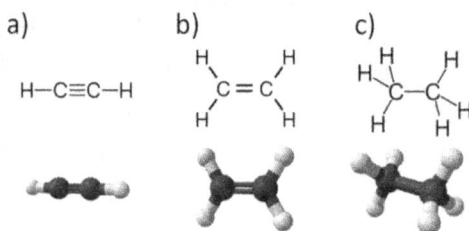

Figure 2.3: *Chemical structure and 3D models of a) acetylene, b) ethylene and c) ethane. There are obvious differences in the geometric alignment of the hybrid orbitals, which is linear for sp-hybridized carbon in acetylene, trigonal for sp2 in ethylene and tetragonal for sp3 in ethane. 3D models taken from reference [25].*

Also, triple-bonds can form in the case of a carbon atom, when it is present in *sp*-hybridization. In this case, the two remaining *p*-orbitals are again perpendicular to the axis of the σ-bond and also perpendicular to each other. This geometry allows the formation of two π-orbitals resulting in a triple-bond together with the σ-bond of the hybrid orbital. Acetylene, a molecule of this class, demonstrates a triple-bond between two *sp*-hybridized carbon atoms, as depicted in Figure 2.3 a).

All molecules, which have been discussed so far in this section, exhibit localized electronic bonds. More precisely, all σ- and π-orbitals are associated with the two atoms forming the bond. But there also exist other types of molecules where the π-bonds are not fully localized and the molecule exhibits more than one

8

possible resonance structure. A very prominent representative showing this phenomenon is the benzene ring, as shown in Figure 2.4.

Figure 2.4: *3D model of the benzene ring with a delocalized π-electron system (left) and two theoretically possible arrangements of single- and double-bonds (middle and right). The σ-bonds between sp^2-hybridized carbon and hydrogen atoms give the hexagonal shape to the benzene ring. The dashed line inside the hexagon (left) indicates the the conjugated π-electron system which is delocalized over the whole ring structure. The alternation of single and double bonds of the two resonant configuration, as shown in the middle and right sketch.*

The ring structure of this molecule is determined by the $σ$ –bonds of the trigonal sp^2 hybrid orbitals. In addition, each of the six sp^2-hybridized carbon atoms contained in the ring provides one $2p$ electron, being available for the formation of a π-bond. In total, three π-bonds are formed along the structure of the ring with two different options for their positioning. If there were only localized bonds, one would expect an alternation of double- and single-bonds between the carbon atoms. Nevertheless, as observed in nature, all present π-orbital electron wave functions delocalize over the whole ring and form a so called delocalized conjugated π-electron system, as shown in Figure 2.4, with bond lengths differing from the ones of single- and double-bonds in localized systems.

In analogy to 1D metals, the semiconducting properties in π-conjugated molecules can be understood in terms of the Peierls instability. According to this, a stoichiometric ratio of 1 between single-and double-bonds in the structure, would provide a half-filled energy band with metallic character to the system. In such a one-dimensional system with incompletely filled bands, the different single- and double bond lengths cause a characteristic lattice distortion and double the elementary cell of the quasi lattice [24]. Therefore, an energy gap E_G between bonding π- and anti-bonding π^*-orbitals occurs, giving semiconducting electronic properties to the conjugated π-electron system. In analogy to inorganic semiconductors, the lower energy band is filled with electrons while the higher energy band remains empty in the ground state. These orbitals, located directly below and above the gap are thus named highest occupied molecular orbital (HOMO) and lowest unoccupied molecular orbital (LUMO). The existence of this energy gap can be revealed by theoretical models, as for example by the theory of Su-Schrieffer-Heeger (SSH-model) [26,27]. Due to the favorable energetic position of the conjugated π-system, the benzene ring, among many other organic conjugated materials, exhibits not only high mechanical robustness but also interesting optoelectronic properties. By controlling the degree of delocalization of the conjugated π-system in more extended systems, this observed energy gap can be tuned on a large scale. Energies for π- π^* transitions down to about 1 eV have been realized, allowing the customization of materials with special optoelectronic properties in terms of light absorption and emission.

Over the past few decades, the possibility of tailoring π-conjugated organic materials to achieve desired optoelectronic properties led to a large number of synthetic organic materials with applications in organic transistors [28], light sensors, light emitting diodes (OLEDs) [7,24,29] and solar cells [30]. The promising idea behind all of these techniques is, to combine the desirable material parameters of plastics, i.e. robustness, mechanical flexibility, light weight and cheap large scale production by roll-to-roll printing, with suitable properties for the above mentioned technical use. A widely-used representative for polymers used in organic photovoltaics is the molecule poly(3-hexylthiophene) (P3HT), as shown in Figure 2.5.

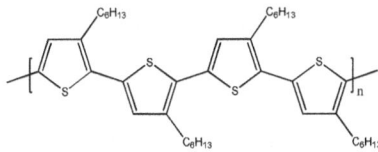

Figure 2.5: *Chemical structure of regio-regular P3HT. Due to the regular alignment of its side-chains, the polymer forms lamellar-like structures in thin films. The resulting inter-molecular interaction plays a beneficial role for light absorption as well as exciton- and charge-transport properties*

In this and many similar materials, the conjugated π-electron system is delocalized over many thiophene rings. The region of which one continuous π-electron system extends is called chromophore, which can spread out over several tens of thiophene rings. A regular alignment of side-chains in regio-regular P3HT leads to the favorable formation of lamellar-like structures in thin films of pristine P3HT. This enhances inter-molecular interaction and the delocalization of the π-electron system over several neighboring chains. This two dimensional delocalization further reduces the ground state energy of the system and plays a crucial role for the onset of light absorption as well as exciton- and charge transport properties [13,31].

2.2. Quantum mechanical description of π-conjugated materials

For a mathematical description of an electronic system and its observables, as for example its electronic structure, energy levels or dynamics, a quantum mechanical approach has to be chosen. All electrons in such a system are described by wave functions Ψ. For a known wave function Ψ it would be possible to calculate the expectation value of any physical observable of the system. However, since it is not possible to precisely determine Ψ and other initial conditions of complex structures, like π-conjugated systems and not even simpler many-electron systems [23], several approximations have to be made to calculate wave functions and to get quantitative estimates of physical observables.

One very important method is the Born-Oppenheimer approximation. This approximation relies on the fact, that the mass of the nuclei in the systems is much larger than the mass of electrons and thus the movement of nuclei is taking place on a much longer timescale than the fast movements of electrons. For this reason, regarding the heavy nuclei as immobile at the timescale of electronic movements and separating their respective movements is a good approximation.

The many-body time-independent Schrödinger-equation for a stationary electronic state, described by the wave function Ψ is given by:

$$\hat{H}\Psi(r) = \left\{ -\frac{\hbar^2}{2m_e}\sum_j \nabla_j^2 - \sum_{j,k} \frac{Ze^2}{|r_j - R_k|} + \frac{1}{2}\sum_{j \neq l} \frac{e^2}{|r_j - r_l|} \right\} \Psi(r) = E\Psi(r) \quad (2.1)$$

Here, $\hat{H}$ represents the Hamilton operator, Z the atomic number, e and m_e charge and mass of an electron, $\hbar$ the Planck constant and E the energy. The variables r_j and R_k denote the coordinates of the electrons and nuclei, respectively. Hereby, bold letters depict vectors. The first term of equation (2.1) describes the kinetic energy of all electrons in the system and the second term is the electrostatic Coulomb interaction between electrons and nuclei. The third term the electrostatic interaction between the electrons.

In general, an exact solution of the Schrödinger equation, even if describing only two electrons, is not solvable in an exact manner [23]. Hence, suitable approximations are needed to reduce the problem to a solvable system, while still providing a physically meaningful picture of the system.

One approximation method is the self-consistent Hartree-Fock method. It is based on the idea of describing the wave function of all N electrons as a product of the N single-electron wave functions, the so called Slater determinant [23]. Here, the electron-electron interaction is simplified, by each electron interacting with an effective potential due to the other remaining electrons. The variation principle thus leads to set of N coupled equations. An iterative, self-consistent calculation allows finding solutions for these N equations and finally provides the molecular orbitals and energies of the system. Nevertheless, the computational effort for these iterative steps limits the number of active electrons in the system, being in many cases not sufficient for a description of extended π-conjugated systems as investigated in this work. Systems of this complexity require large computational effort exceeding justifiable computation time.

A drastic approximation, further simplifying the system, is the so-called Hückel model. In addition to fixed nuclei positions, the Hückel model further neglects any electron-electron interaction [32]. An extension of the Hückel model, including electron-phonon interactions and by this nuclear motion, is the Su-Schrieffer-Heeger model, which was developed especially for describing polymers [26,27]. Strong coupling between electrons and phonons causes an energy gap between the highest occupied molecular orbital (HOMO) and the lowest unoccupied molecular orbital (LUMO).

A more promising approach to calculate the quantity of physical observables with less required computing effort is the Density Functional Theory (DFT). Almost 50 years ago, Hohenberg and Kohn discovered that knowing the ground-state distribution of the electron density $\rho(r)$, as described in equation (2.2), is sufficient to fully describe a stationary electronic system [33–35].

$$\rho(\boldsymbol{r}) = \sum_{j=1}^{N} |\Psi_j(\boldsymbol{r})|^2 \qquad (2.2)$$

In principle, all other observables can be obtained from this quantity. It is a very convenient variable, because the density is a physical observable with an intuitive interpretation, which depends only on three spatial coordinates (instead of $3N$ coordinates like the many-body wave functions of N electrons) [35]. It has been shown that by using a variational principle in terms of density the exact

13

ground-state energy of the system can be attained at the exact density. Furthermore, Kohn and Sham proposed an auxiliary non-interacting system to evaluate the density of an interacting system [33,35]. Within this so-called Kohn-Sham system, the electrons are described by time-independent one-particle Schrödinger equations with an effective external potential V_{KS}, based on a functional of the electron density.

$$\widehat{H_{KS}}\Psi_j(\mathbf{r}) = \left\{-\frac{\hbar^2}{2m_e}\nabla^2 + V_{KS}[\rho](\mathbf{r})\right\}\Psi_j(\mathbf{r}) = E_j\Psi_j(\mathbf{r}) \qquad (2.3)$$

The Kohn-Sham potential V_{KS} consists in three terms,

$$V_{KS}[\rho](\mathbf{r}) = V_{Ext}(\mathbf{r}) + V_{Hartree}[\rho](\mathbf{r}) + V_{XC}[\rho](\mathbf{r}) \qquad (2.4)$$

which are the external potential V_{Ext}, i.e. the Coulomb interaction between the electrons and the nuclei, the classical Hartree part, considering the electron-electron interaction

$$V_{Hartree}[\rho](\mathbf{r}) = \int \frac{\rho(\mathbf{r})}{|\mathbf{r} - \mathbf{r}'|}d^3r' \qquad (2.5)$$

and a so far unknown exchange-correlation potential V_{XC}. The latter potential contains all complex many-body effects and also corrections for the classical treatment of the electron-electron interaction in $V_{Hartree}$. A simple approximation for this functional, based on the exchange energy of a homogeneous electron gas [36], turned out to give quite accurate results for many applications and is still widely used. The use of the electron density instead of many-particle wave functions and the Kohn-Sham system are the basis of the DFT [37]. This approach allows to reduce the many-body problem as multiple single-body problems with a system of independent solutions of one Schrödinger equation.

An extension to DFT, including the interaction of electromagnetic fields with matter, is the time-dependent DFT (TDDFT) [38–40]. Its central theorem proves a one-to-one correspondence between a time-dependent external potential and the electron density for many-body problems for a fixed initial state [40]. This allows to obtain the external potential just by knowing the electron density of

14

the system. With this, now time-dependent external potential, the time-dependent Schrödinger equation can be solved,

$$i\hbar \frac{\delta}{\delta t} \Psi_j(r,t) = \widehat{H_{KS}}(r,t)\Psi_j(r,t) \qquad (2.6)$$

revealing all properties of the system.

During the last decades, TDDFT has become an important tool to calculate electron distributions of ground and excited states of complex organic systems and to give estimations of their energetic levels [41–43].

Figure 2.6 shows a visualization of the electron wave function for the frontier orbitals on a chain of poly(2,6-(4,4-bis(2-ethylhexyl)-4H-cyclopenta[1,2-b;3,4-b']dithiophene)-4,7-benzo[2,1,3]thiadiazole)) (PCPDT-BT), calculated with a TDDFT technique.

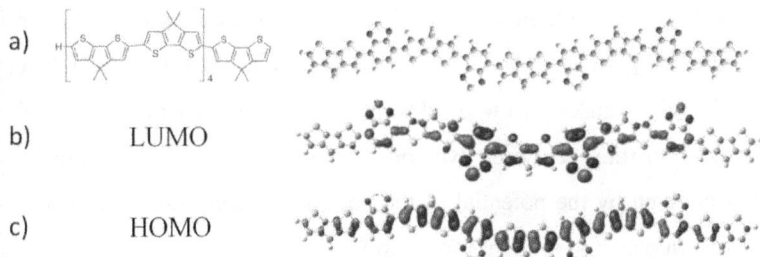

Figure 2.6: a) Chemical structure and graphic illustration of the investigated molecule chain of the copolymer PCPDTBT. The electron wave function distribution of the LUMO is given in panel b) and the HOMO in panel c). Both are calculated with a TDDFT.

Theoretical investigations of this kind are a useful tool to understand physical and optoelectronic properties, like light absorption, and also the temporal evolution of photoexcited states. Not only geometric information of electron wave functions are accessible, but also quantitative parameters like the oscillator strength [44,45]. This information is of utmost importance for the interpretation of experimental spectroscopic results on materials for organic photovoltaics, such as PCPDT-BT.

2.3. Optical properties of π-conjugated materials

The presence of several occupied and unoccupied molecular orbitals in π-conjugated materials allows transitions of electrons between levels of different energy. As explained in section 2.1, the lowest accessible electronic transitions are between binding π- and anti-binding π*-orbitals. In the materials, studied in this work, the required excitation energies are on the order of 1.5 eV, which corresponds to the energy of photons in the visible spectral range. In the following, these orbitals are regarded as HOMO and LUMO, respectively, although in the latter case the property of zero-occupation might be lost when photoexcited.

For the sake of ease, we will describe the physics of light absorption and emission in an excitonic picture, which means we regard a bound electron-hole pair as a quasi particle, called exciton, with different possible energy levels S_n. Figure 2.7 gives an overview of absorption and emission of a photon due to electronic transitions between two orbitals. In each state, the energy of the exciton is given by the potential of the nuclear displacement coordinate R and can persist in one of several vibronic sub-levels, denoted as v.

Because of the energetic spacing between these sub-levels of about 100 meV, only the lowest of them is occupied at room temperature. As can be seen in Figure 2.7, optical transitions correspond to vertical transitions in those diagrams, since electronic movement is taking place on a much faster timescale than nuclear movement due to the large difference in their mass, i.e. $m_e \ll m_n$. This simplification of separating the movements of electrons and nuclei is known as Born-Oppenheimer approximation [46]. Nevertheless, a rearrangement of the electron wave functions inevitably requires new positioning of the nuclei to reach again the energetic minimum [24].

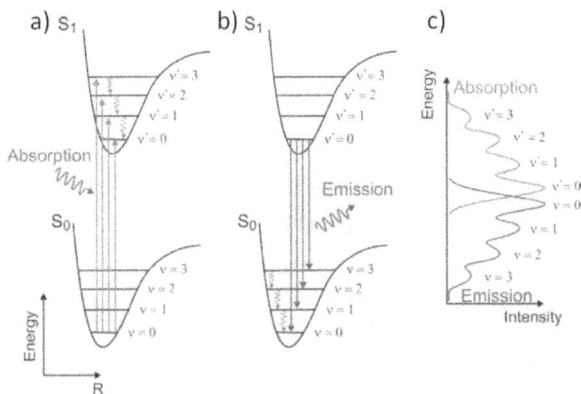

Figure 2.7: Scheme of the electronic energy landscape for two different levels of excitation, i.e. the excitonic ground state S_0 and the first excited state S_1. R denotes the nuclear displacement coordinate and υ numbers different vibronic sublevels. Optical transitions, i.e. a) absorption and b) emission of a photon correspond to vertical transitions (Born-Oppenheimer approximation). Panel c) shows the symmetry of absorption and emission as well as the vibronic progression due to the Franck-Condon principle.

This nuclear displacement causes a horizontal shift of the potential curve of S_0 and S_1, allowing several transitions from the (only) occupied vibriational ground state of S_0 to several vibronic sub-levels in the excited state S_1. The difference between excited state and ground state equilibrium geometries enables a wave function overlap of different vibronic sub-levels. This phenomenon is mathematically expressed as Franck-Condon integrals (FC), which in this case have a non-zero value [47]. From equation (2.7) it is obvious that the absorption cross section σ is directly proportional to the square of the dipole moment of the initial state $S_0(v=0)$, denoted as μ_0^0, and the square of the wave function overlap of initial and final state, termed Franck-Condon integral (FC).

$$\sigma \propto |\mu_0^0|^2 \times |\langle \Psi_1^n | \Psi_0^0 \rangle|^2 = |\mu_0^0|^2 \times FC \qquad (2.7)$$

Due to the orthogonality of these wave functions in the case of identical equilibrium geometries, these Franck-Condon integrals and thus the absorption cross section σ, would vanish and the transitions would be forbidden if no nuclear displacement was present. In this case, only $S_0(v=0)$ - $S_1(v=0)$ would be allowed. The vibronic progression of absorption and emission of light thus originates in distinct equilibrium geometries of different excited states, known as the Franck-Condon principle.

Typically, the emission spectrum is mirror symmetric to the absorption spectrum, as shown in Figure 2.7 c). The reason for the presence of the same vibronic progression in the emission spectrum is that photon emission almost only occurs from the lowest vibronic sub-level of the excited electronic state. The reason is that the relaxation of the exciton from higher-lying to the lowest vibronic sub-levels is taking place on a much faster time scale than the emission of a photon. The first happens typically in only few tens of femtoseconds [48]. The characteristic energy gap between the $S_0(v=0)$ - $S_1(v=0)$ absorption and the $S_1(v=0)$ - $S_0(v=0)$ emission is caused by the reorganization of the molecule in the excited state. This dissipation of energy lowers the potential energy curve and thus the energy of the emitted photon [23].

Up to this point, the spin of electrons and excitons has not been taken into account. There exist two types of excitons with different spins, i.e. singlet states, denoted as S_n, and triplet states, denoted as T_n, with a total spin of 0 and 1, respectively. In Figure 2.8, a Jablonski-diagram shows all relevant transitions in an organic molecule. The excitonic ground state S_0 is a singlet state. For spin conservation rules optical transitions between singlet and triplets are strongly forbidden, and thus optical excitation of π-conjugated materials as they are investigated within this work, mainly lead to the formation of singlet excitons. Only after the formation of singlet excitons, triplet states might be populated inter system crossing between these states. Nevertheless, in π-conjugated

18

polymers for photovoltaics this is typically a very inefficient process, taking place on a timescale of several hundred of picoseconds and nanoseconds [48]. Efficient internal conversion from singlets to triplets and rapid phosphorescence from the triplet state T_1 to the singlet ground state S_0 can be achieved in specially designed molecules, based on organo-metallic complexes [24,49].

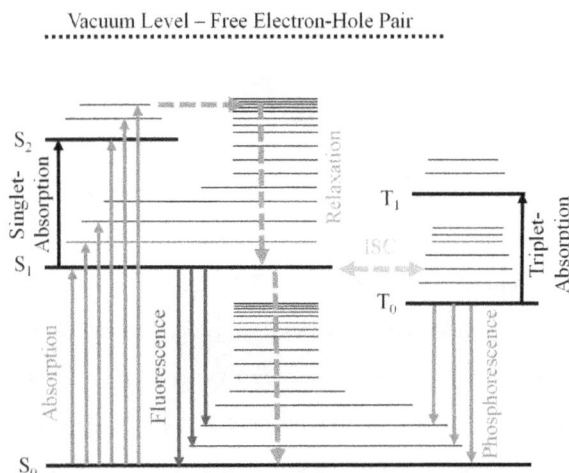

Figure 2.8: Jablonski diagram of excitonic energy levels of a typical organic π-conjugated molecule. Thick and thin black lines represent excitonic energy levels and vibrational sub-levels, respectively. Vertical arrows denote transitions due to light absorption (blue and black), fluorescent (red) and phosphorescent (yellow) emission. Internal conversion, vibrational relaxation and non-radiative deactivation are indicated by dashed green arrows. Inter-system crossing (ISC) accounts for the conversion of singlets to triplets and vice versa.

These molecules incorporate a heavy (compared to carbon atoms) metal atom in the center, e.g. *Pt* or *Ir*, requiring only the total quantum mechanical angular momentum to be conserved, but not the individual spins [24,50]. These interesting small molecule materials find their application mainly in phosphorescent organic light emitting diodes (phOLEDs) [7,50] and differ significantly in their shape and properties from polymers for photovoltaics. For these reasons, the triplet population of the materials studied in this work can be neglected in the further description.

2.4. Photoexcitations in amorphous organic films: excitons and polarons

2.4.1. Excitons

As already mentioned in the previous section, the main species of photoexcitations in organic thin films are so-called excitons [48]. These quasi-particles are not unique for organic materials since they are also known to be present in inorganic semiconductors, but their properties are quite different from those in inorganic systems. They describe in general an electrically neutral pair of electron and hole, bound via Coulomb interaction, and denote quasi-particles allowing a simplified treatment of the interaction with the surrounding matrix of molecules or atoms. In inorganic crystalline materials the exciton binding energy is very moderate and in many cases it is even lower than the thermal energy. For silicon, a binding energy of only about 15 meV has been measured [51], which is well below the thermal energy at room temperature, $k_B T \approx 25$ meV. This leads to quasi unbound electrons and holes upon photoexcitation, with a large excitonic radius of ca. 5 nm. In contrast, excitons in organic materials exhibit a much higher binding energy on the order of 0.3 to 0.5 eV, leading to very strongly bound electron-hole pairs, which are localized typically at only one molecule [52,53]. These are described as Frenkel excitons [48]. Figure 2.9 offers a graphical illustration of the different excitons and their extension in the material system.

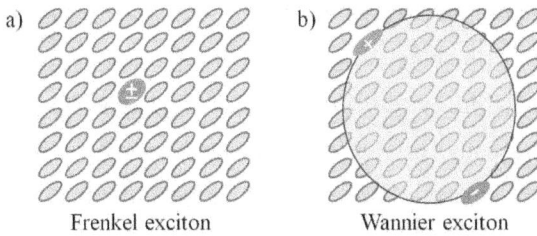

a) Frenkel exciton   b) Wannier exciton

Figure 2.9: *Schematic picture of excitons, i.e. a Coulombically bound neutral electron-hole pair. a) Excitons with a large binding energy and small size, as they are present in organic materials, are described as Frenkel excitons. b) Systems with high electric screening, e.g. inorganic materials, favor the formation of Wannier excitons with low binding energy and large spatial delocalization [48].*

The reason for this pronounced difference in exciton binding energy and extension of organic and inorganic materials is the different electric screening in these materials. While crystalline inorganic semiconductors provide high electric screening expressed by large dielectric constants, i.e. $\varepsilon \approx 11$ for silicon, organic materials suffer from very moderate screening and dielectric constant on the order of only 3 [6,54,55]. Additional contributions to the binding energy are electron-electron and electron-lattice interactions [56]. Consequently, thermal energy at room temperature is by far not sufficient for charge separation. For this reason, strongly bound Frenkel excitons are the main species of photoexcitations in organic materials. Excitons need to overcome a large binding energy (ca. 0.5 eV) in order to dissociate into free charge carriers within their lifetime. Typically, additional electron accepting materials are used in organic photovoltaic devices to ensure efficient charge separation, as explained in detail in section 3.2. Without such acceptors, regular homopolymers only show a small percentage of charge carries, which dissociate mainly at defect sites, impurities [6] or interfaces of different morphological domains with a disordered energetic potential landscape [13,57]. Excitons have been in the focus of scientific investigations during the past years [6,58–64]. Their properties, such as binding energy, mobility, diffusion length and lifetime are known to strongly influence the use of organic materials for various electronic devices. Consequently, exciton migration and energy transport properties of organic materials are a crucial factor for their applicability [65]. An overview of the theoretical models of different exciton transport mechanism will be given in the next sub-section.

2.4.2. Energy transport and exciton migration

A very important property of π-conjugated polymer films is their ability to conduct energy in terms of exciton migration [48]. Energy migration in organic solids can be described as a hopping process from molecule to molecule, unlike a wave-like propagation in crystalline solids, as it is the case in inorganic semiconductors. Due to their electrical neutrality, their motion is not affected by electric fields, but it is governed by a diffusive motion through the system with random directions.

Energy transfer can occur via two nonradiative ways, the Förster- and the Dexter-mechanism. In the Förster-type incoherent energy transfer process, intra- and inter-molecular energy transport is possible and usually acts in downhill direction, to sites with a lower energy [46]. It can end in a trapped state at the low-energy end of the inhomogeneously broadened density of states, often formed by aggregates or defects. Thermal fluctuations are required to overcome the energy barrier of these trap states and to reactivate excitonic motion [6]. During diffusion, eventual vibrational relaxation of the exciton occurs typically on a timescale of less than 100 fs [58]. Förster energy transfer is described by a Coulomb dipole-dipole interaction between a donor molecule (D) and an acceptor molecule (A) [66]. The transfer rate $k_{Förster}$, describing the efficiency of energy transfer is expressed by

$$k_{F\ddot{o}rster} = \frac{3}{2}\kappa^2 k_D^{rad}\left(\frac{R_0}{r_{DA}}\right)^6 \tag{2.8}$$

$$R_0^6 = \frac{3c^4}{4\pi n^4}\int_0^\infty \frac{f_D(\omega)\sigma_A(\omega)}{\omega^4}d\omega \tag{2.9}$$

$$\kappa^2 = \boldsymbol{\mu_D}\cdot\boldsymbol{\mu_A} - 3\left(\boldsymbol{\mu_D}\cdot\frac{\boldsymbol{r_{DA}}}{r_{DA}}\right)\left(\boldsymbol{\mu_A}\cdot\frac{\boldsymbol{r_{DA}}}{r_{DA}}\right) \tag{2.10}$$

where R_0 dentotes the Förster radius and κ an orientation factor between the donor ($\boldsymbol{\mu_D}$) and acceptor ($\boldsymbol{\mu_A}$) dipole moments, given by equations (2.8)(, (2.9), and (2.10). Bold letters indicate variables being vectors.

Here, r_{DA} expresses the distance vector between the donor and acceptor molecule and k_D is the radiative decay rate of the donor. The Förster radius R_0 is the critical transfer distance at which the energy transfer rate is equal to the recombination of the isolated donor molecule [47], and consequently the transfer rate drops to 50%. This transfer rate depends on several constants, including the vacuum light velocity c and the material's refractive index n. A further crucial parameter is the spectral overlap of the normalized donor emission $f_D(\omega)$ and the absorption of the acceptor $\sigma_A(\omega)$. In the case of sufficient spectral overlap between the two participants, which is the case for two identical molecules, Förster transfer is an efficient energy transfer mechanism. Although its efficiency depends strongly on the distance, i.e. $k_{F\ddot{o}rster} \sim r_{DA}^{-6}$, this type of energy transfer shows typically critical distances between $1 - 10$ nm in many organic materials [47,50].

A further way of nonradiative short-scale exciton transfer, which is even more affected by the intermolecular distance, is the above mentioned Dexter mechanism [67]. In 1953, Dexter extended the Förster analysis to a higher multipole-multipole analysis and included electron exchange interactions. At close donor-acceptor distances, before the full quantum mechanical regime begins, quadrupole and even higher terms become relevant for an accurate description [68]. Those multipole interactions decrease rapidly, following higher inverse power of the intermolecular distance and are thus of a shorter range than dipole-dipole interactions. Yet, whenever a dipole-forbidden transition is involved, these contributions may play a significant role. When the donor-acceptor transitions are spin-forbidden, the electron-exchange interaction, described by Dexter, becomes dominant. The

transfer rate k_{Dexter} still depends on the spectral overlap J, but is independent of the optical transition moments [47].

$$k_{Dexter} = BJe^{-\frac{2r_{DA}}{L}} \tag{2.11}$$

The exponential decrease with intermolecular distance originates in the required wave function overlap, which is accounted for the typical penetration depth L of the wave function into the environment. The coupling constant B cannot be related to optical properties. These two mechanisms describe the nonradiative exciton transfer between two molecules. Their critical dependency on the inter-molecular distances explains the large influence of morphology and molecular packing on the exciton mobility in organic films. Typical values for the exciton diffusion length of P3HT are 3 – 9 nm [69,70], with exciton mobilities up to 0.1 cm^2/Vs [28,31]. For the formation of free charge carriers, excitons still have to to overcome the Coulomb attraction of electron and hole. This crucial process, which is one of the first steps in charge carrier formation, shall be discussed in the next section.

2.4.3. Exciton dissociation and polaron pair formation

Not all photogenerated excitons in organic materials share the same fate and undergo intramolecular (geminate) recombination to the ground state, as it was quantitatively described first by Onsager [71]. Although this recombination occurs very efficiently and fast, other exciton deactivation processes exist. One further intramolecular process is the intrinsic charge-carrier formation. The different excited states and their vibronic sublevels of neutral singlet excitons exhibit a permanent configuration interaction with the continuum of states of the ionized molecule. This so called autoionization, with a quantum yield $\phi_0 \ll 1$, is in permanent competition with the much more efficient intramolecular or geminate recombination [48,72], with a complementary quantum yield of $1 - \phi_0$. Figure 2.10 gives a schematic overview on the principle of autoionization. In a first step, absorption of a photon leads to the formation of a singlet exciton in one of the accessible states, i.e. S_1, S_2, S_3 or at even higher energetic states. In the second step, the autoionization occurs. With a quantum yield of ϕ_0, positively charged molecular ions and hot, non-thermalized electrons with remaining kinetic energy are formed. The third step consists in the thermalization of hot electrons, which dissipate the kinetic energy in scattering processes. Due to the Coulomb attraction of the positive molecule ion and the electron, a charge carrier pair is formed [73], where both constituents can either share the same molecule (intra-chain polaron pair) or be separated on adjacent molecules (inter-chain polaron pair) [74].

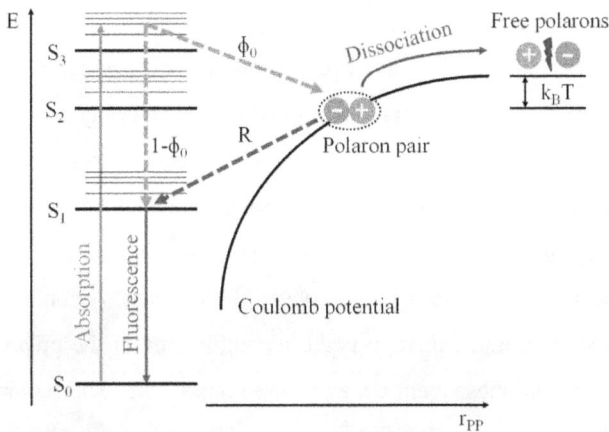

Figure 2.10: *Scheme of the single steps for intrinsic formation of polarons. After absorption of a photon, a formed neutral exciton in a high energetic state can undergo autoionization and form a polaron pair, with a quantum yield ϕ_0, or undergo geminate recombination. Further thermal energy is required to dissociate the weakly bound polaron pairs to free polarons.*

These so-called polaron-pairs have a smaller binding energy than a neutral exciton, because of the reduced Coulomb attraction between positive and negative charge, going along with their further spatial separation r_{PP}. The definition of a polaron and its physical properties will be explained in subsection 2.4.4. Knowing the Coulomb attraction, which still leads to bound charges, a maximum radius for the polaron pair can be estimated, given by the distance r_c, where the Coulomb attraction drops to the same level as the thermal energy persisting in the system. The critical radius r_c for thermal polaron-pair dissociation, also called Onsager radius as it is described in equation (2.12), depends only on one material parameter, i.e. the dielectric constant ε.

$$r_c = \frac{e^2}{4\pi\varepsilon\varepsilon_0 k_B T} \tag{2.12}$$

This demonstrates the strong dependence of charge carrier binding energy on the electric screening and explains the main challenge for charge separation in organic photovoltaic systems, compared to inorganic ones. Regarding equation (2.12), a dielectric constant of 3 leads to a critical radius of r_c of 18.6 nm at a temperature of 300 K. However, the presence of energetic disorder and polarization effects in solid organic semiconductors can easily provoke energetic variations of 0.1 – 0.2 eV [75], which, taken into account together with thermal energy, lead to a more realistic effective Onsager radius of 2.5 – 5 nm.

In the last step of autoionization, the bound polaron pairs are dissociated with a quantum yield of *1-R* by thermally activated dissociation. A competing process for this last step of autoionization is the recombination of the polaron pair to a neutral exciton and subsequent decay to the gound state. The required thermal energy, and thus the critical radius r_c, can be lowered by an additional electric field, overlapping the Coulomb potential. A built-in electric field, as it exists for example between the electrodes of solar cells, can provide such a reduction and assist charge separation. For electron-hole distances smaller than the Onsager radius, an electric fields F leads to a certain probability $P(F)$, for electron and hole to escape their attractive potential, which is given for moderate field strengths by [6,76]:

$$P(F) = exp\left(\frac{-r_c}{r_{PP}}\right)\left(1 + \frac{er_c}{2k_BT}F\right) \tag{2.13}$$

While the probability for this dissociation can be explained by a modified Onsager model [68,71,77], it cannot give answers on the formation process of the polaron pairs. Other theoretical studies by Arkhipov [78,79] and Basko [80] come to the conclusion that excess excitation energy plays an important role for the dissociation of photoexcitations into free charge carriers. The theory by Arkhipov proposes that hot excitons are able to dissociate during the process of thermalization while they still have sufficient excess energies to overcome the Coulomb barrier. This limits the time disposal for dissociation to only few tens of femtoseconds. Similar time estimates are given by Basko, predicting efficient dissociation within the time, in which vibrational degrees of freedom (phonons) are stored in the system after excitation with excess energy. Several experimental observations can be explained with valid time estimates [13,81–83], while for other experiments the predictions remain insufficient [73,84–86]. Hence, the process of this formation in the first 100 fs is still not sufficiently described and understood. For this reason, further theoretical and experimental research is necessary to discover the basic physical processes governing charge carrier formation in π-conjugated materials.

2.4.4. Polaron pairs and polarons

Similar to the strongly bound, electrically neutral electron-hole pair described as Frenkel exciton in the previous sub-section, charge carriers in organic systems are described as quasi particles, called polarons. The lack or excess of an electric charge on a polymer chain leads to the formation of a positive hole-polaron or a negative electron-polaron, respectively. In the case of a short oligomer, they are usually called a radicals. In contrast to free charge carriers in inorganic solids, polarons in amorphous or microcrystalline organic films describe the interaction of the charge with the surrounding molecular matrix. This interaction can be of electrostatic nature or involve electron-phonon coupling [47,87,88]. Due to the interaction with induced dipole moments on polymer chains in the vicinity of the charge, a geometric rearrangement of the molecular backbones leads to the formation of a new equilibrium state at lower energy. Figure 2.11 provides a simple physical picture of charges interacting with the surrounding matrix.

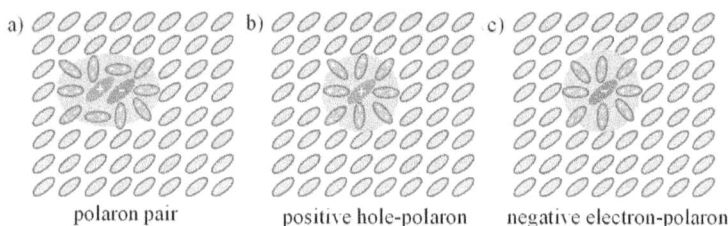

polaron pair   positive hole-polaron   negative electron-polaron

Figure 2.11: *Schematic picture of a) a polaron pair, b) a positive hole-polaron and c) a negative electron-polaron. These quasi-particles describe b) the lack or c) the excess of an electron, together with the resulting electrostatic or electron-phonon interaction with the surrounding molecular matrix. Polaron pairs are bound precursor states of free polarons. All species can be of localized or delocalized nature.*

These polarons may preserve a delocalized nature and extend over several polymer chains, which is observed especially in crystalline organic materials, and behave like a quasi particle with large effective mass [47]. In amorphous solids they are supposed to completely lose their wavelike nature and to localize on one polymer chain or on several adjacent chains. In this case, the resulting geometrical relaxation is regarded to be restricted to the surrounding molecules. Polarons carry spin ½ and consequently are doublets with allowed values of ±½ for the spin orientation.

In addition, there exists a precursor state of these quasi free charge carriers in amorphous organic films, which is the species of weakly bound polaron pairs, as they were already mentioned in the description of exciton dissociation. This weakly bound state describes a quasi particle, consisting of a negative and a positive polarons, suffering still from weak Coulomb attraction.

Because of the small dielectric constant, the occurrence of polaron pairs is very typical of amorphous organic solids. Polaron pairs exist as an intermediate step between full ionization of the involved molecules and recombination of the polarons to an exciton or the electronic ground state. The spatial separation of the involved hole- and electron-polaron leads to a negligible wave function overlap. For this reason, the physical properties of isolated polarons remain largely unaffected and the properties of polaron pairs can be regarded as the sum of its constituents [47]. This has important consequences for the interpretation of spectroscopic measurements, as it will be discussed in chapter 5.

Geometrical relaxation and the rearrangement of the molecule into a new ground state configuration lead to the formation of new electronic energy levels, which are distinct from the neutral molecule. For the same reason, also the optical bandgap, which is given by the π-π* transition of the neutral molecule, is lowered by the presence of intra-gap states [88] (compare section 5.2). The difference of distortion energy E_{dis}, which is necessary for the rearrangement of the polymer chain, and the energy shift ΔE of the border orbitals (down-shift of LUMO and up-shift of HOMO) defines the binding energy between polaron and matrix [88]:

$$E_{PM} = \Delta E - E_{dis} \qquad (2.14)$$

Experimental values for E_{PM} in P3HT and other homopolymers are reported to be 30 – 60 meV [28].

Because of the great importance of electrically conductive organic polymers for various electronic applications, like optoelectronics, photonics and photovoltaics, the properties of electric charges in such systems have been widely studied during the past years [28,88–101]. Most of the properties, like binding energy, mobility or lifetime, are strongly affected by material parameters and cannot be generalized for all π-conjugated material classes. Therefore, the scope of this work is to extend the knowledge about charge carriers to a new class of materials, which in most cases up to now has been restricted to homopolymers and oligomers. This new class of materials is low-bandgap copolymers, which are discussed in the following chapters.

3. Background and state of the art of organic photovoltaics

This chapter provides an overview to the state of the art and the limitations of photovoltaics in general. It will be followed by a detailed discussion about the working principle of organic solar cells, and different approaches in their design. Finally, low-bandgap copolymers are introduced. This novel material class for organic photovoltaics, which is used in most recent and record-breaking organic solar cells, will be discussed in detail. Their unique chemical structure, which is based on an alternated arrangement of donor and acceptor segments, will be presented, together with its beneficial properties in terms of a reduced optical bandgap.

3.1. State of the art and limitations of photovoltaics

The vast majority of solar cells on the market are single junction silicon devices known collectively as first generation devices [102]. Since their invention more than 50 years ago at the Bell Labs in New Jersey, they are still based on the use of crystalline silicon for their active layer with a typical thickness of hundreds of micrometers. More recently, development has focused on the fabrication of thin-film solar cells with active layer thicknesses of only about 1 µm. Although these second generation devices show lower efficiencies than thick first-generation devices, their thinner active layer allows a reduction of material and fabrication costs and also reduces the requirements of charge carrier diffusion length and material crystallinity [103]. For these reasons, other materials were used, including amorphous or micro-crystalline silicon, cadmium-telluride (CdTe) or copper-indium-gallium-diselenide (CIGS). Nevertheless, although the required amount of raw materials is strongly reduced, the disadvantage of using rare and toxic materials in these systems cannot be neglected.

Both generations suffer from the disadvantage of a fundamental limit of their efficiency of about 30% due to thermodynamic reasons [102]. This so-called detailed balance limit of efficiency of p-n junction solar cells was found in 1961 and is often named after its discoverers, William Shockley and Hans Queisser [104]. In their analysis, they assume that the sun and the solar cell are black-body radiators with respective temperatures T_S and T_C. As an ultimat efficiency, $\eta_{ultimate}$, they assume a perfectly absorbing photovoltaic device, leading to one generated electric charge for each absorbed photon of the frequency v, carrying an energy, which exceeds the optical bandgap $E_G=h\upsilon_G$ of the solar cell.

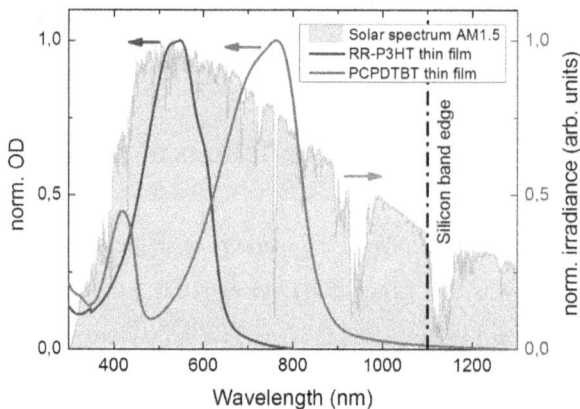

Figure 3.1: *Spectrum of solar irradiance (grey) under AM1.5 conditions, taken from NREL [105]. The blue dashed line depicts the absorption onset of silicon. The red and blue curves show the absorptivity of thin films of two commonly used polymers for photovoltaics, i.e. RR-P3HT (red) and PCPDTBT (green), respectively. Polymers offer large flexibility for tuning the optical bandgap and the onset of absorption.*

The maximum achievable ultimate efficiency is theoretically reached when each quantum of the Planck distribution of the solar spectrum with an energy exceeding E_G results in a charge carrier carrying the elementary charge e and simultaneously an open circuit voltage equal to the bandgap $V_{OC}=V_G=E_G/e$ are achieved. For a solar temperature of 6000 K, the maximum efficiency is reached with a bandgap of 1.1 eV, as it is present in silicon [104].

Taking into account geometrical considerations regarding the incident sunlight and imperfect light absorption, as well as radiative and non-radiative loss mechanisms, leads to further restrictions for the upper limit of solar cells. This so-called nominal efficiency is given by the product of V_{OC} and the short circuit current I_{SC}, divided by the incident power P_{in}, as shown in equation (3.1).

$$\eta_{nominal} = \frac{V_{OC} \cdot I_{SC}}{P_{in}} = \eta_{ultimate}(T_S, E_g) \cdot \varphi \cdot n(f, T_S, T_C, E_g) \qquad (3.1)$$

I_{SC} originates in the removal of holes and electrons from the p-doped and n-doped region, respectively. In equation (3.1), φ denotes the probability of a photon with sufficient energy getting absorbed and $n(f, T_S, T_C, E_g)$ considers geometrical issues and loss processes[104].

Further, the actually achieved maximum efficiency for a real solar cell is derived by the maximum rectangular, fitting below the current voltage curve. It defines the maximum values for voltage V_m and current I_m, as shown in Figure 3.2.

They define the so-called fill factor (FF), being a measure of the efficiency for the removal of electrons and holes from the n- and p-doped region, respectively.

$$FF = \frac{P_m}{V_{OC} \cdot I_{SC}} = \frac{V_m \cdot I_m(V_m)}{V_{OC} \cdot I_{SC}} \qquad (3.2)$$

Finally, the actual maximum power conversion efficiency is given by the detailed balance limit of efficiency η

$$\eta = \frac{P_{out}}{P_{in}} = \frac{V_m \cdot I_m(V_m)}{P_{in}} =$$

$$= \eta_{ultimate}(T_S, E_g) \cdot \varphi \cdot n(f, T_S, T_C, E_g) \cdot FF(f, T_S, T_C, E_g) = FF \cdot \frac{V_{OC} \cdot I_{SC}}{P_{in}}$$

(3.3)

In conclusion, the parameters FF, I_{SC} and V_{OC} appear as crucial parameters determining the efficiency of solar cells. Assuming a perfect silicon solar cell with E_G = 1.1 eV, T_S = 6000 K and T_C = 300K under normally incident solar light, with perfect light absorption and radiative recombination as the only loss mechanism, yields a maximum efficiency of η = 30% [104], known as Shockely-Queisser-Limit.

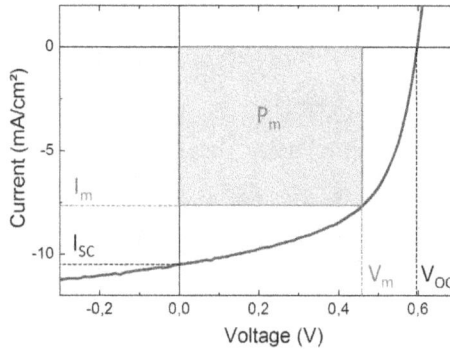

Figure 3.2: *Current-voltage (I-V) curve of an organic solar cell under irradiation, based on a bulk heterojunction of the low-bandgap copolymer PCPDTBT and PCBM (1:2). The maximum power P_m is given by the product of maximum current I_m and voltage V_m. V_{OC} and I_{SC} denote short-circuit current and open-circuit voltage, respectively. Device fabricated and measured by D. Riedel.*

Typically, the efficiencies of real solar cells are well below 30%. This deviation from the ideal case is due to several losses, leading to significant deviations from V_{OC} and I_{SC} and a low fill factor. The current flow in a realistic solar cell can be expressed by a sum of different contributions

$$I = I_{SC} + I_0\left[exp\left(\frac{e(V - R_S I)}{n k_B T_C}\right) - 1\right] + \frac{U - R_S I}{R_P} \tag{3.4}$$

where I_0 denotes the reverse-bias saturation current and $1 \leq n \leq 2$ the diode ideality factor. R_S represents a serial Ohmic resistance originating in bulk, contact and circuit resistance of solar cell devices. Loss processes besides radiative losses are accounted for in the parallel resistance R_P.

Figure 3.3 sketches an electric circuit being representative for a solar cell [106]. In the limit of an infinitely high R_P and vanishing R_S, the case of an ideal solar cell is described. Altogether, these effects reduce the today achievable efficiency in real single-layer devices to about 20% in inorganic and 10% in organic solar cell devices.

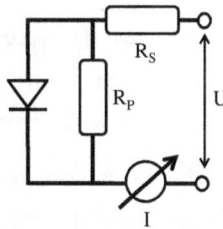

Figure 3.3: *Equivalent circuit of a solar cell, containing a diode, Ohmic resistances R_P and R_S, voltage and current. The parallel resistance R_P accounts for non-radiative losses in competition with charge carrier formation, the serial resistance R_S represents bulk-, contact- and circuit- resistance.*

Newest developments and research, regarded as 3rd generation photovoltaics, aim to circumvent the limitations of the Shockley-Queisser-Limit, using different approaches. Among them are multi-junction cells [107], which harvest photons of different energies in different layers with a reduced loss of photon excess energy, as well as hot carrier cells, aiming to extract charge carriers maintaining their excess energy [108]. Also multi-exciton generation is a promising attempt to generate several charge carriers by the absorption of only one high-energy photon [102,109,110]. One part of this new generation is also organic photovoltaics [30]. It includes dye-sensitized solar cells [111], hybrid solar cells, which consists in inorganic nanoparticles dispersed into a semiconducting polymer matrix [112,113] or in inorganic nanostructured semiconductor templates filled with organic semiconductos [114], as well as all-organic approaches. The latter, especially polymer solar cells, will be discussed in more detail in the next section.

3.2. Working principle of organic solar cells

In addition to wet electrochemical solar cells, all-organic solar cells have become a highly popular research topic during the last decade. All-organic techniques are divided into several areas [30], i.e. small molecule solar cells deposited from gas phase [115–118] and solution processed solar cells. The latter ones include (i) all-polymer solar cells [119], (ii) small moluecule solar cells [120,121] and (iii) polymer/fullerene based solar cells, which are the most important division. All materials investigated in this work are specially designed molecules for polymer/fullerene solar cells; therefore the further description of the working principle will focus on this device class. This class has experienced tremendous progress in the past few years, and breaks the barrier of 10 % power conversion efficiency in the year 2012, as demonstrated by *Heliatek GmbH* [122].

Many of the most efficient organic solar cells are based on blends of highly absorbing polymers, optimized for light absorption and hole conduction, with an electron accepting material, such as fullerenes or inorganic materials [30,123]. Without loss of generality, we will further consider only [6,6]-phenyl-C_{61}-butyric acid methyl ester (PCBM) [124] as electron acceptor. It is a commonly used derivative of the Buckminsterfullerene C_{60} [125]. These constituents form the active layer for light absorption and charge carrier formation of the solar cell

[4]. The strong binding energy of excitons requires them to reach an interface with acceptor molecules as a prerequisite for efficient exciton dissociation, as discussed in section 2.4 [6]. Limited diffusion lengths thus often require good mixing of the polymer with electron accepting materials. To ensure a large interfacial area between donor and acceptor materials, different strategies have been developed, ranging from a simple bilayer structure of thin donor- and acceptor films in planar heterojunctions [116,126,127], over an inter-penetrating network of a mixed blend in bulk heterojunctions [128–130], to sophisticated micro-structured assemblies in interdigitated heterojunctions [131–133]. Figure 3.4 gives an overview of the layout of common polymer/fullerene solar cells [134] and typical heterojunctions of donor and acceptor materials in the active layer.

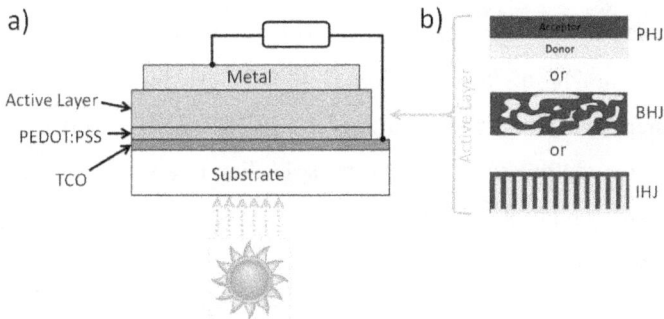

Figure 3.4: *a) Layout of a polymer/fullerene solar cell, consisting in a flexible substrate, a transparent conducting oxide (TCO) as bottom contact, an electron-blocking layer (PEDOT:PSS), the active layer and a metal top-electrode. b) In the active layer, donor and acceptor materials can be arranged as planar heterojunction (PHJ), bulk hetero junction (BHJ) or via micro-structuring as interdigitated heterojunction (IHJ).*

Each blend layout has several advantages and disadvantages of exciton dissociation and charge carrier extraction. While BHJs allow in most cases efficient exciton dissociation due to a large interfacial area, the extraction of holes via polymer channels and electrons via fullerene channels might be limited due to interrupted pathways towards the electrodes. On one hand, PHJ might provide perfect conditions for charge carrier extraction but on the other hand it can provide only limited interfacial area. Thus the dissociation site might be out of reach for diffusing excitons. An attempt to merge both advantages is the realization of micro-structured interdigitated systems [133], but their fabrication often turns out to be sophisticated, time-consuming and expensive.

Once an exciton reaches an interface of donor and acceptor molecules via thermal diffusion, the formation of a charge transfer (CT) state takes place with an efficiency of almost unity [123,135]. This high efficiency can be attributed to a significant driving force for charge carrier separation, which is realized by pronounced LUMO level offsets between donor and acceptor [6], as depicted in Figure 3.5. Needless to say, this electron transfer from the donor to the acceptor LUMO goes with a significant loss of voltage, reducing the reachable open circuit voltage V_{OC} of the solar cell [136–139]. Although theoretical studies predict already efficient charge transfer for a LUMO offset of only 0.1 eV, many systems are based on much higher offsets, leading to a significant reduction of the maximum achievable power due to a linear relationship between V_{OC} and the LUMO offset [137].

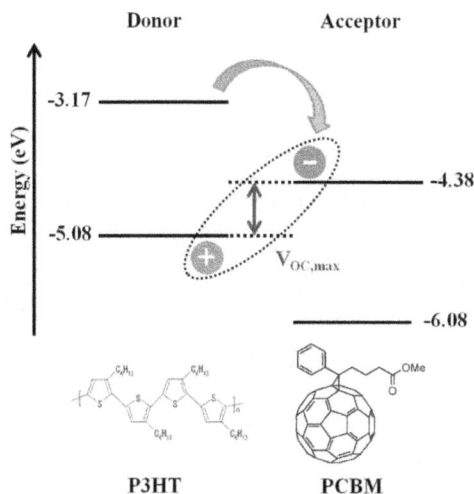

Figure 3.5: *Scheme of relevant energy levels for charge separation at a donor-acceptor interface. Energy values and chemical structures are given for P3HT and PCBM as donor and acceptor material, respectively. A large LUMO offset ensures efficient charge transfer but limits also the theoretically achievable $V_{OC,max}$.*

Extraction of the separated and electrically charged carriers is driven by the built-in voltage, originating in different work functions of the used electrode materials [140] and the chemical potential gradient within the cell [141]. A typical choice of electrode materials for these purposes is a conducting transparent oxide, for instance indium tin oxide (*ITO*), as a transparent bottom electrode, allowing light to penetrate into the cell, and a reflective top aluminum electrode [4]. The obtained built in voltage is typically in the order of 0.3 – 0.5 V for commonly chosen materials, and results in very substantial electric field strength through the device, because of its thickness of typically only few hundreds of nanometers. An additional layer of poly-ethylene-dioxythiophene:polystyrenesulfonic acid (PEDOT:PSS) increases the work function of the positive electrode and acts as an electron blocking layer. Such multilayer structures can be fabricated with a thickness of less than 1 μm, excluding the substrate and the encapsulation. These are required to prevent penetration of oxygen and moisture. The ability to wet-process the organic materials allows roll-to-roll printing and a promises a high potential for large scale production [4]. In order to be feasible for large scale production and an extended commercial use, the obtained solar cell efficiency has to be further increased. A very important step in this direction has been made by using novel materials, i.e. low-bandgap copolymers. These materials are in the focus of the next section.

3.3. Novel materials for organic solar cells: Low-bandgap copolymers

The reachable efficiency of conjugated polymer solar cells critically depends on the material choice for the active layer. Due to the weak absorption of the commonly used PCBM as an electron acceptor material, the light harvesting properties are mainly given by the used polymer. As explained in section 3.1, the optical bandgap of the polymer, and consequently of the resulting solar cell, crucially influences the coverage of the solar spectrum and the achievable power conversion efficiency. When using wide-spread homopolymers like poly(2-methoxyl-5-(3,7-dimethyloctyloxy)-para-phenylene-vinylene) (MDMO-PPV) or P3HT as electron-rich light absorbers in combination with PCBM, the attainable power conversion efficiency is limited to about 2.5% [142,143] and 5% [144–146], respectively. This can be partially explained by the absorption, covering the solar spectrum in an insufficient way, as shown in Figure 3.1.

To overcome this limitation, a novel class of materials has been developed in the past years with the aim to flexibly control optoelectronic parameters, like HOMO- and LUMO-energy levels and the optical bandgap. The synthesis of copolymers, consisting of an alternating assembly of electron-rich and electron-deficient moieties along the molecule backbone, allowed the control of these parameters [30]. Figure 3.6 a) shows a sketch of the alignment of electron-donating and electron-accepting segments along the molecular backbone, given the name donor-acceptor copolymers or push-pull materials.

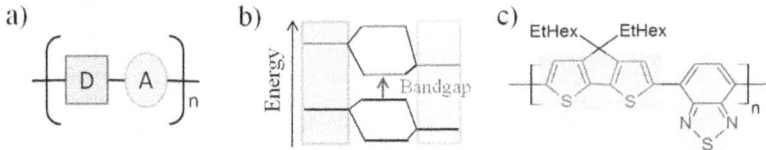

Figure 3.6: *a) Scheme of the alternating alignment of donor- (D) and acceptor-moieties (A) along the molecular backbone forming a copolymer. b) Schematic visualization of the energy levels of isolated donor and acceptor segments, forming new energy levels with a reduced optical bandgap due to electronic coupling. c) Chemical structure of the low-bandgap copolymer PCPDTBT with marked donor- (blue) and acceptor (yellow) regions.*

Due to the electronic coupling between these two moieties, as depicted in Figure 3.6 b), new energy levels form, with a reduced optical bandgap. As an example, in Figure 3.6 c) the chemical structure of of PCPDTBT is shown, with an electron donating dithiophene and an electron accepting benzothiadiazole highlighted in blue and yellow, respectively. Figure 3.1 clearly illustrates the lower optical bandgap of PCPDTBT compared to P3HT, with its characteristic double-peaked absorption, called "camel-back" spectrum [147]. For sake of clarity, it should be mentioned that this donor-acceptor copolymer approach has neither the intention nor the capability to substitute the addition of acceptor materials like PCBM in a photovoltaic blend. The focus of this approach lies in the reduction of the optical bandgap. For efficient charge separation, a heterojunction with electron-accepting materials is still needed. A large variety of such low-bandgap materials has been synthesized and used in devices, based on many different kinds of donor moieties, like polyfluorene, polycarbazole, or cyclopentadithiophene, and acceptor moieties, such as benzothiadiazol or diehexylthienopyrazine [19,123,148–154]. The possibility of shifting the HOMO and LUMO energy levels separately from each other by a suitable choice of donor and acceptor moieties allows not only adjustment of the absorption onset, but also an optimization of the energy level offset between copolymer and fullerene. To some extent, this gives the option to increase V_{OC} by matching the LUMO of the copolymer and the used acceptor material [30]. The high power conversion efficiencies achieved with low-bandgap copolymers, compared to cells based on homopolymers, leaves no doubt that the enhanced infrared is benefitial for photovoltaic application. So far, power conversion efficiencies of more than 10% have been obtained, exceeding by far those of cells with homopolymers [122,123]. Also with photovoltaic devices based on short donor-acceptor oligomers, a power conversion efficiency as high as 6.7% has been demonstrated recently [155].

Beside their reduced optical bandgap and the high efficiencies of solar cells using these novel materials, only little is currently known about the optoelectronic properties of these materials. Since the validity of the theories describing homopolymers are not expected to fit for the decription of these copolymers, a detailed study of photoexcitations in these systems is presented in the following part of this work, with a special focus on the formation of weakly bound polaron pairs, i.e. the precursor state of free charge carriers.

4. Experimental and theoretical methods

In this chapter a detailed description of the main experimental and theoretical techniques will be given, which have been used to conduct the physical investigations presented in this work. Insights into the basic working principles, the experimental realization as well as the technical capabilities and limitations are presented. The described techniques include the fabrication of thin film samples of organic materials, steady state spectroscopy on electrically charged molecules and different techniques for ultrafast femtosecond spectroscopy. Last but not least, a brief summary of used theoretical methods will complete the chapter.

4.1. Fabrication of thin organic polymer films

For investigating the optical and physical properties of organic molecules, thorough preparation of suitable sample devices is of utmost importance. Especially ultrafast pump-probe spectroscopy measurements require the availability of thin polymer films with controlled properties, attached to a non-disturbing substrate. To ensure good transparency of the substrate in a wide spectral range, which is required from the blue at 400 nm down to the MIR at 5 μm wavelength, substrates made out of CaF_2 with a thickness of 1 mm were chosen. This material provides sufficient broad transparency starting at ca. 190 nm and exceeding 7.7 μm.

To allow the preparation of polymer films with a thickness on the order of only 100 nm with good optical quality and to prevent their contamination with any kind of impurity, all used substrates underwent a careful cleaning procedure. This procedure consists of an iterative cleaning of the substrates with isopropanol and acetone in an ultrasonic bath for ten minutes with two repetitions, each. Finally, drying them under a nitrogen stream assured to obtain cleaned substrates without any impurity.

Preparation of thin polymer films on a clean substrate was achieved by spin coating. Figure 4.1 shows a sketch of the working principle of the spin coating technique. For spin coating, a chosen cleaned substrate is mounted on a rotary table, where it is fixed by low pressure from a vacuum pump

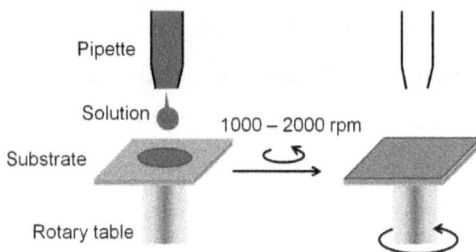

Figure 4.1: *Working principle of the spin coating technique. A cleaned substrate is fixed with low pressure to a rotary table. The desired amount of polymer solution is dropped from a pipette onto the substrate which is then rotated around the axis perpendicular to its surface, spreading the solution in a homogeneous fashion. Rotating the substrate during the whole drying process of the solvent leaves behind a homogeneous thin film of the desired polymer.*

Directly after placing a drop of the polymer solution at the center of the substrate on the rotary table, it is mechanically rotated around the axis perpendicular to the sample's surface. This rotation causes a homogeneous spread of the polymer solution on the whole substrate surface. Rotation for three minutes turned out to be sufficient for the evaporation of the whole solvent, leaving behind a thin and homogeneous film. The desired polymer was solved in a suitable ultrapure solvent, for which typically toluene was chosen. Hereby the polymer concentration in solution plays a key role for the resulting film thickness. Typically the thickness has been adjusted to values between 5 mg/mL and 20 mg/mL, which are limited at the upper end by the solubility of the polymer. Heating to 75° C and stirring the solutions for several hours ensured that the polymer was well dilute in the solvent. Not only the concentration, but also the rotation speed has a critical influence on the homogeneity of the resulting film and its thickness. With values between 1000 and 2000 rounds per minute (rpm), together with the above mentioned concentrations, film thicknesses between 50 to 200 nm were obtained. Variation of both parameters allowed adjusting the thickness and hence the optical density of the polymer films to the respective spectroscopic needs. The latter typically was been kept between 0.1 and 0.6 for the majority of the studied samples by pump probe spectroscopy, which was achieved with concentrations between 5 mg/mL and 10 mg/mL and a rotation speed of 1800 rpm for the studied donor-acceptor copolymers.

4.2. Femtosecond pump-probe spectroscopy

In the past decades a very powerful technique has been elaborated, which allows studying molecular electronic processes and molecular reactions at their specific timescale. The experimental approach towards measurements on a timescale as short as 10^{-15}-10^{-12} seconds, i.e. femtoseconds, thus was called femtochemistry, and its discovery was rewarded with the Nobel Prize in the year 1999 to Ahmed Zewail [156]. Since then further development of femtosecond spectroscopy did not stop but has undergone an impressive evolution towards a shorter time resolution and a broader spectral coverage, ranging from the ultra-violet down to the THz frequency range. Today, cutting edge technology uses femtosecond laser systems and nonlinear optical techniques, such as optical parametric amplifiers (OPAs) [157], allows measuring molecular temporal dynamics with a resolution as short as 20 fs [158]. The basic idea of femtosecond pump probe spectroscopy is depicted in Figure 4.2.

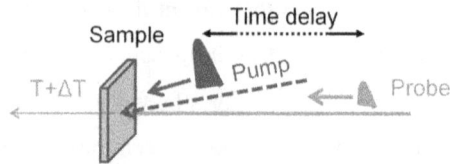

Figure 4.2: *Working principle of femtosecond pump-probe measurements. The transmission T of a low-intense spectral broad- or narrowband probe pulse through a thin film sample of the desired material is measured. Upon photoexcitation of the same sample spot with a pump pulse at a desired excitation wavelength before every second probe pulse, this transmission is altered due the presence of photoexcitations. The relative change in transmission ΔT/T is measured for various pump-probe time delay steps on femto- and picoseconds timescale.*

It is based on detecting the transmission T of low-intense probe pulses (typically with repetition rates from kHz to MHz) through a thin film sample of the desired material on a transparent substrate. Details about the preparation of suitable films can be found in section 4.1. For this purpose, the probe pulse can consist in a broad spectrum or in a narrow band spectrum, at a suitable wavelength range to cover the spectroscopic signatures of interest. By focusing an additional high-intense pump pulse of the desired excitation wavelength to the same position on the sample, where the probe pulse is transmitted, a certain fraction of molecules is photoexcited and leads to a change in the transmittance $\Delta T(\lambda)$ at the respective spectral positions of the probe pulse. By mechanically changing the optical path length of one of the pulses and consequently the delay between pump and probe, the temporal evolution of the relative change in transmission $\Delta T/T$ can be measured on a femto- and picosecond timescale. Depending on the material and the energetic position of the spectral features of interest, different setups have to be chosen for obtaining the best results. In the following sub sections, two different setups will be explained. The first one has been used for time-resolved measurements of polaron absorption in the MIR while the second one allows spectral broadband measurements in the visible and NIR, tracking ground state bleaching, stimulated emission and the onset of polaron absorption at the same time. For a detailed discussion of the measured spectral features of photoexcitations in conjugated materials the reader is referred to chapter 5.

4.2.1. Ultrafast spectroscopy in the near- and mid-infrared spectral region

For femtosecond pump-probe measurements in the NIR and MIR a home-built setup has been used [44]. The system layout with the main components is depicted in Figure 4.3.

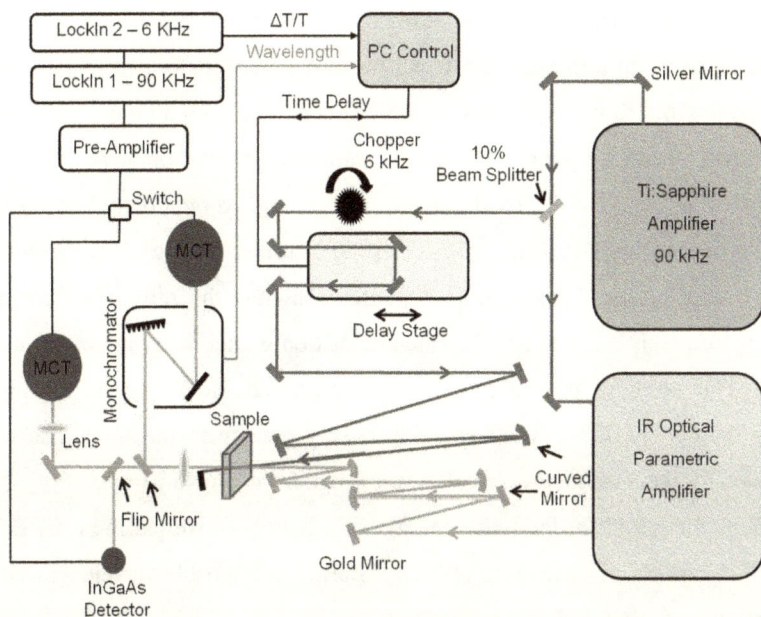

Figure 4.3: Schematic layout of the used home-built infrared pump-probe setup. A Ti:sapphire amplifier provides pulses at 800 nm for excitation of the sample and frequency conversion to the IR in an OPA to probe wavelengths from 0.95 to 5 µm. The transmitted probe intensity is detected by InGaAs or MCT diodes

and amplified by lock-in amplifiers. This setup allows sensitive measurements with a $\Delta T/T$ detection limit of $3 \cdot 10^{-4}$.

The setup is based on a Ti:sapphire regenerative amplifier (RegA 9060, COHERENT), emitting 60 fs light pulses (FWHM) with a pulse energy of 6 µJ at a central wavelength of 800 nm and a repetition rate of 90 kHz. About 10% of the pulse energy is split off with a beam-splitter and further used as excitation pulse for the pump-probe experiment. Eventual focusing to a nonlinear optical β-barium-borate (BBO) crystal and subsequent recollimation allows frequency doubling [159] for experiments with 400 nm excitation wavelength. Mechanical chopping of this pump beam reduces the repetition rate to 6 kHz (15 pulses on, 15 pulses off). The further path guides the pump beam over a mechanical translation stage, which is capable of inserting a delay between the pump and the probe pulse up to a duration of 1 ns. The last mirror, before the pump pulse reaches the thin film sample, has a spherical curvature with a radius of 600 mm in order to focus the pump beam to the sample. Typically obtained focal spot diameters are on the order of 200 µm and are measured on a daily basis with a CCD beam profiler (WinCamD, LASER2000). Further folding of the beam path before reaching the focusing mirror assures obtaining the zero-delay temporal overlap of pump and probe at the sample position within the range of the delay stage. This accounts for the compensation of the additional pass of the probe pulse through IR OPA. The excitation energy density in the focus has been adjusted for the most measurements to the range from 7 to 45 µJ/cm^2.

The remaining 90% of the pulse energy is frequency converted in an IR OPA, which is capable of generating ultrashort pulses in a tunable wavelength range from 0.95 to 5 µm [160]. For wavelength longers than 1.4 µm, careful dispersion management is possible using the bulk dispersion of different materials. While the bulk group delay dispersion (GDD) of Si and Ge is positive in the whole investigated wavelength range and can be used for up-chirping (delay of blue vs. red wavelengths) the probe pulses, the GDD of CaF$_2$ becomes negative at wavelengths longer than ca. 1.4 µm. This allows to compensate for the dispersion of optical components in the OPA and of Si and Ge windows. the latter ones are used as spectral filters after the OPA for blocking the fundamental laser radiation at a wavelength of 800 nm. Adding bulk CaF_2 windows of various thicknesses to the optical beam path induces a down-chirp (delay of red vs. blue wavelengths) and allows optimizing the probe pulse duration by adjusting the second order dispersion close to zero. This fabulous and simple technique for pulse compression is exclusively available in the infrared spectral range, due to the absence of materials with negative GDD in the visible and ultraviolet wavelength range [161]. By measuring a crosscorrelation [162] between 800 nm pump and 3000 nm probe pulse with a specially designed nonlinear optical $LiNbO_3$ crystal (200 µm thickness, cutting angle $\theta = 30$), a width of only 122 fs (FWHM) could be measured for the instrument response function (IRF), as shown in Figure 6.4 b).

To ensure that the whole probe beam is affected by the photoinduced changes to the investigated material in the sample, the probe focus needs to be smaller in diameter than the spot irradiated by the pump beam. Because of the large difference in wavelength, which is up to a factor of five longer for the probe, the same focal lengths and beam diameter lead also to an up to five times larger focus, as known from the Abbe diffraction limit. To ship around this limitation, a much shorter focusing length had to be chosen for the probe, i.e. 12.5 mm. By using a telescope with spherical mirrors, the beam diameter has been increased by a factor of two before focusing, to shrink the achievable focal size. Typically values for the probe beam focus of ca. 100 µm could be achieved, which are of about half the diameter of the pump beam. In general, the use of gold mirrors in the whole optical path of the probe beam reduced the losses in reflectivity through the entire used infrared spectral range.

After transmission through the sample, the remaining energy of the pump pulse gets dumped by a beam block. The transmitted probe beam is recollimated by a

CaF$_2$ lens, providing good transmission at the used IR wavelengths. The transmitted probe intensity T is detected after focusing, either by an InGaAs- or a nitrogen cooled mercury-cadmium-telluride-diode (MCT) (MCT-5-N, INFRARED ASSOCIATES). These detectors are sensitive in a wavelength range from 0.8 – 1.7 µm and ca. 1.5 – 6 µm, respectively. Depending on the used probe wavelength, one of these detectors can be selected by deflecting the beam with a flip mirror or letting it pass through. Alternatively, another MCT detector with a preceding monochromator (MICRO-HR, HORIBA JOBIN YVON) can be chosen to optimize the probe spectrum in the IR beyond the sensitivity range of InGaAs-diode array based spectrometers. However, spectrally resolved pump-probe measurements are not carried out with this setup. Spectral resolution in this wavelength range would not contribute to a gain in knowledge. The broad spectral width of polaron signatures located at IR wavelengths, which typically cannot be fully covered by the probe spectrum generated by the OPA. The generated voltage signals in the detectors undergo a first preamplification by three orders of magnitude (MCT1000, INFRARED ASSOCIATES) and are further read by two lock-in amplifiers (SR830DSP, STANFORD RESEARCH). The preamplified signal is fed into the first lock-in amplifier, together with a trigger signal synchronized with the 90 kHz laser repetition rate, measures a voltage signal U_{90}, which is proportional to the average transmitted probe intensity impinging at the detector, i.e. $U_{90} \sim T + \Delta T/2$. For this measurement, rather weak bandpass filters (6 dB/octave) and short time constants (typically 10 or 30 µs) must be chosen, in order not to cut off the frequency region carrying the signal of the fourier transformed temporal modulation with 6 kHz. The amplitude of this modulation originates in photoexcitations generated by the pump pulses, which occur at an envelope repetition rate following the chopping frequency 6 kHz. This corresponds to the ΔT signal we are interested in, and can be detected either at the difference or the sum of both frequencies, which is 84 and 96 kHz, respectively. By feeding the output signal of the first lock-in amplifier into the second one, the trigger frequency of 6 kHz of the chopper controller can be used for detecting the signal at 96 kHz. Because the first lock-in passes through the 90 kHz signal as a DC voltage, it renormalizes the frequency spectrum. In a figurative sense this can be understood as a subtraction of 90 kHz, bringing the original 96 kHz signal to the more convenient frequency of 6 kHz. Typically, the bandpass filters and the time constants of the second lock-in could be chosen in a very effective manner, i.e. 24 dB/octave and 300 ms, in order to get a clean signal at the desired frequency. The amplitude of the voltage signal measured by the second lock-in corresponds to $U_6 \sim \Delta T/2$. Consequently the desired information can be obtained by reading the output of both lock-in amplifiers (U_{90} and U_6) with a computer and evaluating it by:

$$\frac{\Delta T}{T} = \frac{2 \cdot \frac{\Delta T}{2}}{\left(T + \frac{\Delta T}{2}\right) - \frac{\Delta T}{2}} = \frac{2 \cdot U_6}{U_{90} - U_6} \qquad (4.1)$$

This signal is recorded for every desired delay step, together with the corresponding wavelength information. Delay steps down to 10 fs were found to be convenient for measuring ultrafast processes on picoseconds time scales, while large time steps were chosen to track temporal dynamics on a time scale of several hundreds of picoseconds. The readout of the lock-in amplifiers and the positioning of the mechanical delay stage have been realized by the controlling computer, run by a homebrewed software.

In total the presented home-built IR pump-probe setup allows to measure femtosecond spectroscopy with excitation wavelengths of 400 and 800 nm and a tunable probe wavelength in the range from 0.95 to 5 µm. A temporal resolution of 122 fs was achieved and measured by cross correlation between pump and probe pulse. The high repetition rate of 90 kHz with a modulation of the pump pulses at a chopping frequency of 6 kHz, allows a high sensitivity with a background noise level of only $5 \cdot 10^{-5}$. Relative changes in the probe transmission ($\Delta T/T$) of only $3 \cdot 10^{-4}$ can measured successfully, what demonstrate the powerful capability of this setup. All in all, this is a very suitable setup to detect polaron absorption also for very moderate excitation densities and under exclusion of any bimolecular recombination effects [44].

4.2.2. Ultrafast spectroscopy with broadband probe covering the visible and near-infrared

Ultrafast spectroscopic measurements with probe wavelengths in mid-infrared allow measuring the polaron absorption band P_2 of many materials without any overlapping spectral features of other photoexcitations [44]. For this purpose the experimental setup introduced in the previous section has been applied successfully. Nevertheless, for other measurements it might be useful to observe the spectral features of polarons, excitons and ground state bleaching within one measurement. This can be done with another approach in ultrafast spectroscopy, using spectrally broadband white light for probing. The latter covers the entire visible and near-infrared range [13,163]. For this purpose, a home-built setup at the Center for Ultrafast Science and Biomedical Optics in Milan has been used. This setup is based on a Ti:sapphire regenerative amplifier (INTEGRA C, QUANTRONIX), delivering 100 fs pulses at a central wavelength of 800 nm, with 1 mJ pulse energy at a repetition rate of 1 kHz. In principle, the apparatus is designed in a very similar way as the setup presented in Figure 4.3. Nevertheless, there are slight differences, accounting for generation of the different pump and probe wavelengths, which are discussed in this section.

Spectrally tunable excitation pulses are generated by a single stage OPA, pumped at 400 nm. It allows tuning the excitation wavelength in the range from ca. 500 to 800 nm. High energy excitation, with wavelengths in the blue and UV as short as 370 nm, is obtained by optional subsequent frequency doubling of the OPA output. For probing, a supercontinuum white light is generated by focusing about 2 μJ of fundamental pump laser light at 800 nm onto a sapphire plate. The resulting spectral broadband white light can be optimized towards good coverage of the visible spectral region below 800 nm. Alternatively, it can be optimized toward the NIR, by choosing suitable focusing and a sapphire plate of 2 or 3 mm thickness, respectively [164]. This white light pulse is not temporally compressed but shows a chirp (delay of blue relative to red wavelengths) and has a pulse-duration of few picoseconds, which is encountered during the data extraction and evaluation. To benefit from this broadband white light probe pulse, a detector is required allowing to spectrally resolve the probe. For this purpose a combination of a spectrograph with a silicon CCD array (Stresing Entwicklungsbüro) was chosen [165]. This system is capable of detecting selected broadband wavelength ranges from 350 nm up to 1100 nm. Due to distortions of the spectral phase at the wavelength of thefundamental laser light at 800 nm, and different requirements of the alignment for the white light generation, separate measurements for the spectral ranges below and above 800 nm have to be carried out. This configuration allows recording 2D maps of the obtained signal $\Delta T(\lambda,t)/T$, with spectral resolution on one axis (λ) and the temporal pump-probe delay at the second axis (t). The temporal chirp due to the stretched white light pulse has to be carefully taken into account and can be removed by using homebrewed software for the evaluation. A big benefit of this 2D map is that it offers to extract the recorded signal spectrum $\Delta T(\lambda)/T$ at any recorded time delay and to extract the temporal decay $\Delta T(t)/T$ at any desired wavelength within the spectrum. In this way, spectral shifts, originating in the eventual overlap of spectral features of different species of photoexcitations, can be identified by investigating the temporal evolution of the signal spectrum. With this technique a temporal resolution of ca. 150 fs was achieved, which is comparable to the one the of IR pump-probe setup, as detailed in the previous section. Because of the relatively low repetition rate of 1 kHz, which allows the beneficial spectrally resolved detection, a sacrifice of detection sensitivity has to be made. The achieved sensitivity allows detecting signals down to $5 \cdot 10^{-4}$. Although the excitation densitiy has been kept always as moderate as possible, bimolecular effects could not be excluded in every measurement. This was especially the case, when measuring polaron absorption bands, which are relatively weak compared to ground state bleaching or stimulated emission.

In conclusion, this pump-probe setup allows to record 2D $\Delta T(\lambda,t)/T$ maps and offers high-level spectral and temporal resolution within the same measurement. The broad probe spectrum, covering the visible and near-infrared make it a very powerful tool for observing spectral features of different photoexcitation species within the same measurement. A free choice of the excitation wavelength allows to adress the investigated materials at their specific wavelengths and to study the role of excess excitation energy on polaron formation in conjugated polymers and oligomers, as presented in chapter 7 and in reference [166].

4.3. Steady state spectroscopy on charged molecules

Measuring formation, presence and decay of polarons in conjugated organic materials by optical spectroscopy requires a detailed knowledge about the spectral features of the involved ions. For retrieving these spectral features and an unambiguous identification of cations (hole polarons) and anions (electron polarons), a variety of techniques has been used, measuring the differential absorption in a steady state fashion, i.e. without temporal resolution. These techniques allow to some extent to retrieve not only the qualitative evaluation of polaronic spectroscopic signals but partially also quantitative numbers. The latter describe their absorption behavior, i.e. the molar extinction coefficient ε and the polaron absorption cross section σ, as explained in section 5.2.4. This chapter gives an overview of the used experimental techniques that are required to retrieve the differential absorption of polarons in conjugated polymers, and points to the limitations of certain techniques. The discussion will focus to the chemical oxidation, the attempted chemical reduction and the injection of charges by charge modulation spectroscopy.

4.3.1. Cation generation by chemical oxidation with SbCl₅

Over the past two decades several techniques established to measure the cation absorption spectrum of various conjugated molecules by chemical doping either in solution or in thin films. Especially oligo(phenylenevinylene)s as well as oligomers and polymers based on thiophene were studied, by doping with SbCl$_5$ and nitrosonium hexafluorophosphate (NOPF$_6$) in solution [91,167], and with FeCl$_3$ and fluoroalkyl trichlorosilane (FTS) in thin films [168,169]. All these measurements consistently reported the formation of new intra-bandgap absorption bands of the polymers or oligomers, evolving upon doping with the respective oxidant. In this section, the technique of chemical oxidation with SbCl$_5$ and its further use to retrieve the absorption spectra of cations (hole polarons) in donor-acceptor copolymers is explained [44]. As an oxidant, the molecule SbCl$_5$ has been chosen, since it is well known to be a very strong oxidant and to create hole-polarons in most organic materials [170]. A further benefit of this molecule is its spherical geometric arrangement, as shown in Figure 4.4, which promises a reduction of geometrical dependencies of the oxidation process during the interaction with polymer chains in solution.

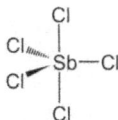

Figure 4.4: Chemical structure of SbCl$_5$. This strong oxidant ionizes most organic materials. Geometrical dependencies of the interaction with polymer chains are expected to be strongly reduced due to its spherical arrangement.

For measuring the differential absorption of conjugated materials, dilute solutions of all investigated materials are prepared with a concentration of 15 $\mu g/mL$ solved in 1,2-dichlorobenzene (ODCB) (CHROMASOLV, SIGMA ALDRICH). The absorption of these solutions is measured with an automated absorption spectrometer (CARY 5000, VARIAN) using fused silica cells

(HELLMA and SPECTROCELL) with a light pass length of 10 mm. With this combination the spectral range from 175 to 3300 nm can be covered, with restrictions below 290 nm and between 2560 and 2860 nm due to absorption of the fused silica cells and the solvent. To prevent the occurrence of absorption by the solvent, various solvents have been tried. ODCB turned out to be the best compromise, because other solvents like toluene showed more pronounced absorption features in the spectral range of interest. However, solvents without hindering absorption features tended to react with the dopant, such as tetrachloroethylene. After measuring the absorption of pristine solutions, consecutive doping with small amounts of a dilute $SbCl_5$ solution at a concentration of 90 µg/mL is carried out. Adding 5 µL of this solution to 3 mL of polymer solution corresponds to a doping ratio of 1% of the dopant relative to the weight of the polymer contained in the cell. Repeated absorption measurements reveal the changes between consecutive doping measurements, as shown in Figure 4.5 for the copolymer PCPDT-BT.

Upon increasing the doping ratio, the ground state absorption of the molecules decreases and prominent absorption bands P_1 and P_2 arise in the infrared due to polaron absorption, as depicted in panel a). For a better comparability, the differential absorption is plotted in panel b). It depicts the absorption of the respective doped solution after subtracting the one of the undoped solution. In this detailed view, it becomes obvious that the response of the spectroscopic feature does no longer grow linear but starts reaching saturation at doping concentrations higher than 4%. Thus, for quantitative measurements it is strongly recommended to stay at low doping concentrations, not higher than 4%, in order to justify the assumption that every inserted dopant molecule creates one hole polaron on a polymer chain.

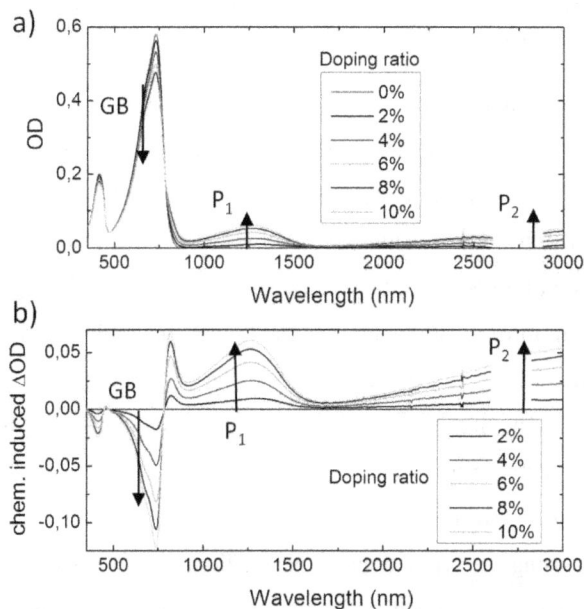

Figure 4.5: *a) Absorption spectra of PCPDT-BT in solution, undoped and doped with various concentrations of SbCl$_5$. b) Chemically induced differential absorption. For increasing doping ratios growing spectroscopic signatures of ground state bleaching and polaron absorption bands P$_1$ and P$_2$ emerge. Above a ratio of 4% the signal grows in a sub-linear fashion due to an onset of saturation.*

From these measurements a qualitative absorption spectrum of cations can be retrieved, which allows to choose the right probe wavelength for their detection by ultrafast spectroscopic measurements. In addition, quantitative numbers for the absorption cross section of polarons can be extracted, as discussed in detail in section 5.2.4.

4.3.2. Anion generation by chemical ionization with strong reducing agents

Similar to the chemical generation of cations in various conjugated polymers and oligomers, we tried to detect the induced change in optical density by chemical reduction of the molecules in solution. Recently, it has been shown that upon reduction with a sodium-potassium alloy (NaK_2) anions of 2,7-(9,9-dihexylfluorene) oligomers can be generated [171]. In a similar way, by using a solid potassium mirror, oligomers of the homo-material phenylenevinylene could be reduced successfully, which showed very similar absorption behavior of cations and anions in these oligomers [91,100]. By doping with the reducing agent diisobutylaluminium hydride (DIBAH), various kinds of copper chalcogenide nanocrystals could be reduced [172].

Both methods, using the liquid dopants NaK_2 (liquid alloy at room temperature) or DIBAH at strongly diluted concentrations in the solvent tetrahydrofuran (THF), have been applied to reduce the investigated donor-acceptor copolymers. The copolymers were solved in the same solvent at a concentration of 15 µg/mL. For DIBAH a concentration of 45 µg/mL allowed to obtain small doping ratios of few percent, similar to the oxidation experiments with $SbCl_5$. Due to poor solubility of the NaK_2 alloy in THF, an exact concentration could not be determined. A comparison of the differential absorption of PCPDT-BT for both dopants can be found in Figure 4.6. Both experiments result in strong ground state bleaching. In the case of hydride as the used dopant, an additional spectral feature from electro absorption (*EA*) occurs in PCPDT-BT. Electro absorption originiates in a shift of ground state absorption due to induction of electric dipole moments on the polymer chain, when interacting with surrounding molecules or external electric fields. The same absorption band has been found for PCPDT-BT when chemically oxidized, as shown in Figure 5.7 b). It has been also reported for other polymers, such as P3HT [96]. Nevertheless, doping with both reduction agents did not result in the expected formation of polaron absorption bands P_1 and P_2.

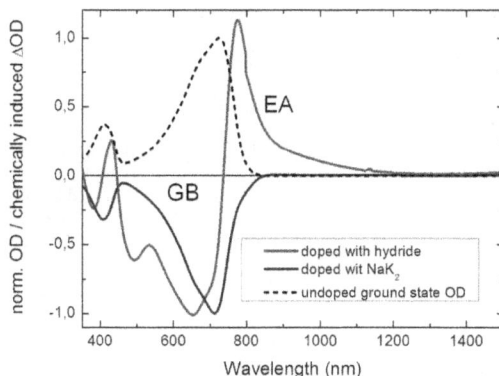

Figure 4.6: *Differential absorption spectra of PCPDT-BT in solution upon chemical reduction with NaK$_2$ (blue) and 10% hydride (red). For comparison, the absorption of the undoped solution is shown as a dashed black line. While doping with alkali metals only results in ground state bleaching (GB), doping with hydride leads to well pronounced electro absoprtion (EA).*

This unexpected outcome for both reduction experiments lead to the assumption, that donor-acceptor copolymers and in general thiophene polymers suffer from poor stability upon reduction. One further indication supporting this assumption is that to our knowledge so far no successful chemical reduction is reported for thiophene polymers or oligomers. It is worth mentioning that successful chemical reduction of fluorene oligomers is reported in literature as mentioned before [171], which is so far the only investigated material in the scope of this work, which could not be oxidized (as polymer polyfluorene) with the dopant SbCl$_5$. A possible explanation for this finding is that the relative arrangement of energy levels of thiophene and donor-acceptor copolymers allows efficient and stable oxidation, while in contrast PFO allows stable reduction when using strong reactants like alkali metals.

4.3.3. Electron injection by charge modulation spectroscopy

A further attempt to retrieve the absorption spectrum of electron polarons in donor-acceptor copolymers has been undertaken by using charge modulation spectroscopy (CMS) [44,173]. Unipolar charge injection can be achieved and thus electron-only devices realized by a suitable choice of electrode materials. These devices exclusively inject electrons to the polymer and are thus a suitable method for measuring its anion absorption spectrum [174]. To ensure good electron injection qualities, metals with high work functions have been chosen. Aluminum with a thickness of only 13 nm has been evaporated as bottom electrode, which resulted in a transparency of ca. 50% through the whole visible and NIR spectrum. The top of this electrode was oxidized for several samples by exposing it to air for several minutes. As detailed in reference [174], the resulting Al_2O_3 layer creates a potential barrier for electron injection but not for extraction. Nevertheless, no significant influence of the oxid layer on the CMS sample performance could be found in these experiments. Afterwards, a thin layer of the desired copolymer was spin coated on top of the bottom electrode. Slight n-doping of the polymer film with dimethylcobaltocene (2% wt.) (DMC) helps filling the trap states and results in a higher electron mobility of the investigated polymer [14,175,176]. Carrying out the spin coating and evaporation steps under inert gas atmosphere prevented the sample from degradation due to oxidation or the intrusion of moisture. On top of the polymer layer the top electrode for electron injection is evaporated under ultra high vacuum (10^{-6} mbar), consisting of 5 nm Ca with a high work function and finished by a thick layer of evaporated Al, acting as a reflecting back electrode. A detailed sketch showing all different layers of a typical electron-only CMS sample is presented in Figure 4.7 b). Above, in Figure 4.7 a), a layout plan of the optical setup for charge modulation setup can be found.

As a broadband light source, a Xe-short-arc lamp (LS0308, L.O.T. ORIEL) has been used, covering the visible and NIR spectral region for probing the *OD* of the CMS sample during charge injection. A monochromator (MICRO-HR, HORIBA JOBIN YVON) provides selectivity of a narrow band wavelength, which is further focused onto the CMS sample. After transmission of the incident beam through the glass substrate and the semitransparent bottom-electrode it is reflected at the top electrode and sent to a detector after proper recollimation. With a silicon- and an InGaAs-diode the whole visible and NIR spectral range can be covered, up to a wavelength of about 1700 nm. The actual charge injection in the copolymer film is achieved by applying a voltage between top and bottom-electrode with a function generator. Voltage between the electrodes of 4.8 V / 0 V for the on / off cycle were chosen. The voltage signal followed a rectangular step function at 320 Hz repetition rate. Lower repetition rates down to 30 Hz did not alter the obtained signal. Figure 4.8 compares the CMS signal obtained for the copolymer PCPDT-BDT, doped with 2% of DMC in an electron only device fabricated with the layout shown in Figure 4.7 b).

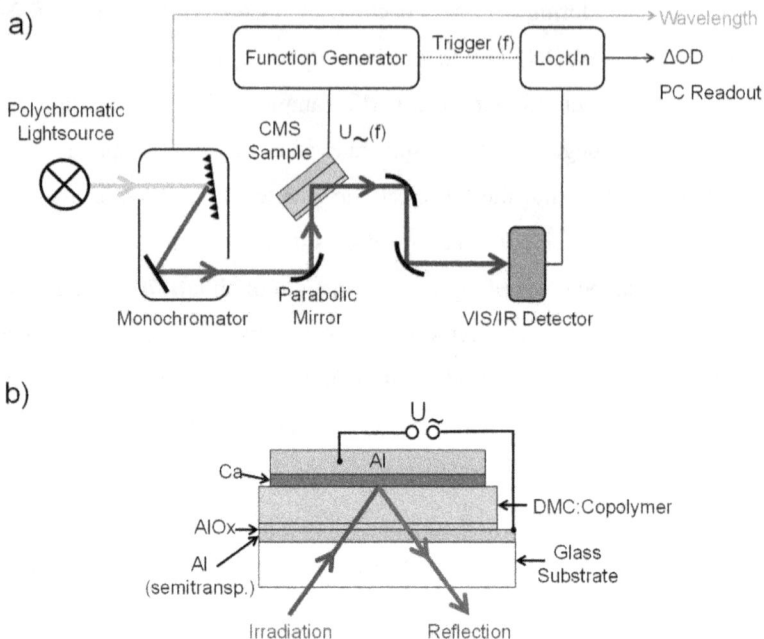

Figure 4.7: Scheme of setup and sample cell for charge modulation spectroscopy (CMS). a) Optical setup for measuring CMS, consisting in a light source, covering the visible and NIR specral region, a monochromator for wavelength selectivity, and a silicon or InGaAs diode for measuring the light after reflection in the sample cell. A function generator applies a voltage between the two electrodes and injects charge carriers. The alternating OD of the sample due to the injected charge carriers is detected by a lock-in amplifier, reading the voltage of the target diode together with the wavelength, adjusted in the monocromator. b) Cross section of a CMS electron-only sample cell, consisting of a semitransparent Al electrode (13 nm) with an oxidized layer, a layer of copolymer (100 nm) doped with DMC (2% wt) and an evaporated top electrode of Ca (5 nm) and thick Al (120 nm) for good reflectivity.

Interestingly, it shares a very similar spectral shape with the cation absorption spectrum obtained by oxidizing the copolymer in solution with $SbCl_5$. Up to

1300 nm, a prominent ground state bleaching can be observed which is then overtaken by the onset of the P1 absorption band. This measurement demonstrates that the principle, setup and device of charge modulation spectroscopy works for polymers, as it was shown also in other experiments [31].

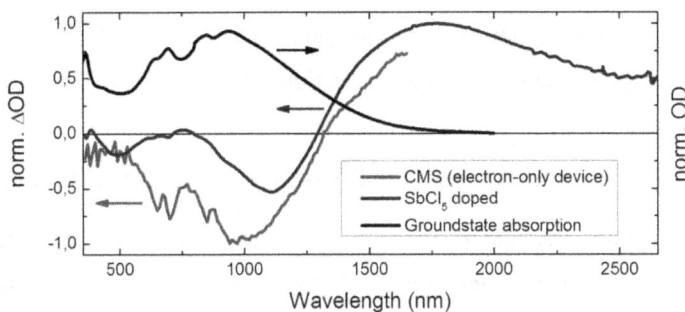

Figure 4.8: *CMS signal (red line) of an electron-only device of PCPDT-BDT doped with DMC (2% wt) and the cation absorption spectrum measured by SbCl$_5$ doping (blue line). Similar to the cation signal, the anion spectrum shows ground state bleaching (coinciding with ground state absorption, black line) and the onset of the polaron absorption band P$_1$ starting at 1300 nm. The limited spectral range of the InGaAs detector allowed to measure the CMS signal only up to 1700 nm.*

However, out of the investigated donor-acceptor copolymers, studied in this work, only for PCPDT-BDT a reliable CMS signal could be obtained. The fact that only this copolymer, exhibiting the lowest optical bandgap and the lowest LUMO energy level, leads to a working device might point to charge injection problems. A mismatch between the work function of Ca and the energy levels of the copolymers might be the origin for the poor performance of CMS samples, when using other donor-acceptor copolymers. Similar to the attempts of retrieving the anion absorption spectrum by chemical doping with reducing agents, also its retrieval by CMS remains a demanding task. For this reason these experiments shall be regarded as a proof of principle. First, they show that it is possible to obtain the anion absorption spectrum by CMS on electron-only devices with certain demands towards the energetic position of the copolymer's LUMO level. Second, it turns out that the P_1 absorption band of cations and anions share a similar, but yet not identical spectral shape. Because of the great importance of knowing the polaron absorption spectra for the studied materials in a precise fashion, theoretical investigations have been performed, to clarify the remaining questions, as detailed in section 5.2.2.

4.3.4. Theoretical methods for quantum chemical modeling

For the quantum chemical modeling of the molecular properties, e.g. absorption spectra and natural transition orbitals (NTOs), two different methods have been used. The goal of the first method was a detailed theoretical investigation of spectral polaron signatures in donor-acceptor materials, revealing the asymmetry between electron and hole wave functions. Furthermore, NTOs of neutral molecules for different excitation energies and for positively or negatively charge ions were calculated. The results of this investigation can be found in sections 5.2.1 and 7.3 and are published in references [45] and [166], respectively. The calculations are based on density-functional theory (DFT) and linear response time-dependent DFT (TDDFT), and were done in the Gaussian09 program suite [177]. The Coulomb-attenuating method Becke three-parameter Lee-Yang-Parr hybrid functional (CAM-B3LYP) and 6-31G* basis set were employed. The visualization was done using GaussView 5 [178]. To capture the long range character in the electronic exchange interaction, which is required to describe the expected pronounced charge-transfer character in the electronic transitions, the long-range corrected CAM-B3LYP functional was used [179,180]. For mimicing the properties of the corresponding polymers, calculations were done for long but finite-length oligomers, consisting of three to four repeating units. To reduce the computational costs, alkyl side chains were replaced by methyl side chains, what is justified by their lacking involvement in optical excitations. Only small quantitative differences were found when comparing the spectra of smaller oligomers with and without symmetry constraints. The dependence of the spectra on the length of the oligomers was studied and no qualitative changes were found. To achieve better comparability, the total chain length was chosen to be similar for all oligomers compared in this work. The results for the oligomers with more repeating units, i.e. PCPDT and PCPDT-BT were obtained using C_{2V} symmetry, and C_1 symmetry for PCPDT-2TBT. Also for absent symmetry constraints in the calculations, only small changes in the molecular structures were observed upon adding or removing charges to the PCPDT and PCPDT-BT oligomers. A relaxation toward slightly more planar structures could be found for PCPDT-2TBT, which is similar for cation and anion. For all calculations, the molecular geometry in the ground state was optimized first and then the lowest 20 electronic excitations were calculated. To retrieve the shown electronic spectra, a homogeneous broadening of 100 meV was applied. Different molecular-orbital contributions to each transition were inspected to obtain a qualitative picture of electronic excitations in the studied systems. The quantitative analysis of the character of electronic transitions was based on the introduction of natural transition orbitals. These cast each electronic transition in a minimum number of pairs of effective single-particle orbitals. While in an ideal case only one relevant NTO pair remains for each transition, a larger number of NTOs is typically found if correlations play a significant role in the electronic excitation.

Because of the demanding computational effort required for the TDDFT calculations, a different approach was chosen for the comparison of the cation and anion absorption spectra of a larger number of different donor-acceptor copolymers, as presented in section 6.4 and published in [44]. For this approach, the semi-empirical Austin Model 1 (AM1) was chosen, which is based on the Hartree-Fock method [181]. With this method, first the ground-state geometry of the studied oligomers has been optimized for an increasing size up to four repeating units. Afterwards, the charged molecules have been calculated by combining AM1 with a configuration interaction scheme [181]. Also for this approach oligomers of similar physical length were chosen, allowing direct comparability between the different materials. The resulting lengths are four repeating units for the copolymers PCPDT-BDT and PCPDT-BT and three repeating units for PCPDT-2TBT and PCPDT-2TTP. As a further simplification, the ethyl-hexyl side chains have been replaced by hydrogen atoms since they do not contribute to the lowest optical transitions. The resulting absorption spectra have been obtained on the basis of the electronic excited states by coupling the AM1 model to a full configuration interaction scheme involving 20 occupied and 20 unoccupied energy levels. As expected, the calculated absorption spectra are shifted toward higher energies, compared to the experimentally obtained ones of the significantly longer copolymers. Nevertheless, also in this case the chosen length is sufficient to mimic the photophysical behavior of the corresponding copolymer.

In total, both chosen theoretical methods gave helpful insights into the photophysical properties of donor-acceptor copolymers. Their combination with state of the art experimental techniques thus played a key role in the discovery and explanation of polaron pair formation in these novel materials.

5. Spectroscopic signatures of excited states in organic materials

Spectroscopic investigation of photoexcitations of various species always requires signatures, acting as individual finger prints for proofing their existence. This can be either photoinduced absorption (PIA), photoluminescence (PL), or photobleaching, which is accompanying each species of excited state. In order to address these features with optical spectroscopy, the spectral region of their occurrence has to be accessible to a suitable optical experimental setup. Today, a wide spectral range can be covered with different spectroscopic techniques, ranging from the ultraviolet (UV) [182,183], through the visible spectral region [157,165,184] down to the mid infrared (MIR) [160,185,186]. Many π-conjugated materials show absorption features in the UV and visible spectral region [20] and photoluminescence at visible (VIS) and near infrared (NIR) wavelengths. For this reason, optical spectroscopy is a widely used tool to study electron dynamics in organic materials, often with ultrafast time resolution by the use of femtosecond pump-probe spectroscopy. In the following sections, an overview of spectral signatures of neutral excitons and polarons in homopolymers and low-bandgap copolymers will be given, with a special attention on the differences of these two material classes and consequences for the physical interpretation of such measurements.

5.1. Exciton absorption and photoluminescence

The presence of excitons in π-conjugated materials can be measured qualitatively and quantitatively with several spectroscopic techniques. One prominent spectroscopic signature of excitons in conjugated polymers is excited state absorption (*EX*). Excitons at the energy level S_1 can be further optically excited to occupy higher energetic states S_n, as depicted in Figure 5.1 a). The required photon energies for these transitions in several polymers are typically in the NIR [61,62,98,187,188] and vary according to the ground state absorption with conjugation length and energetic disorder in the material. Oligomers with short conjugation lengths and therefore blue-shifted absorption allow the detection of *EX* together with polaron absorption and ground state bleaching within the same spectral window of the spectroscopic measurement, ranging from the UV to the NIR [166].

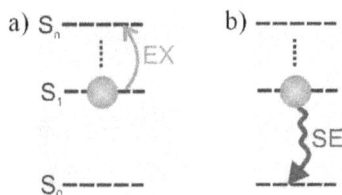

Figure 5.1: Scheme of the energetic levels of an electron-hole pair bound as an exciton. Typical spectroscopic signatures of excitons are the excited state absorption (EX) from S_1 to a higher energetic state, shown in panel a) and the stimulated emission (SE) due to radiative decay (fluorescence) as shown in panel b).

Additional experimental proof for the presence of excitons can be obtained by observing stimulated emission, as shown in Figure 5.1 b). Following Kasha's rule, as explained in section 2.3, stimulated emission and the related relaxation of the molecule to the ground state always occurs after preceding vibrational relaxation from a higher to the lowest vibronic level. Since in the materials studied in this work, photoexcited states are originally all singlet excitons and optical singlet-triplet transitions are strongly forbidden, stimulated emission consists almost only in fluorescence. It is an important spectroscopic signature, allowing to study the lifetime of excitons and geminate (mono-molecular) as well as non-geminate (bi-molecular) decay processes, even on an ultrafast timescale [6,13,44,135,166,187–191]. Because of the large wave function overlap of different exciton states S_0 - S_n, the *EX* and also the *SE* band show very large oscillator strength and consequently large absorption cross sections. This makes optical spectroscopy a very efficient technique for probing the presence of excitons in π-conjugated materials.

5.2. Polaron absorption

5.2.1. Symmetric polaron signatures in homopolymers

Qualitative and quantitative spectroscopic investigation of the presence of charge carriers is of great importance for the understanding and characterization of π-conjugated materials, especially in terms of their application in electronic or photovoltaic devices [28,98]. Polarons provide spectral features in the NIR and MIR spectral region. As described in detail in subsection 2.4.4, the presence of an excess charge as well as the lack of a charge, leads to a geometric realignment of the polymer chain. An accompanying shift of the ground state energy leads to the formation of new optical transitions within the optical band-gap of the neutral molecule, which appear for most polymers in the NIR and MIR spectral range [28,88–101]. Figure 5.2 gives a schematic overview of the optical transitions of polarons in homopolymers and radicals of oligomers.

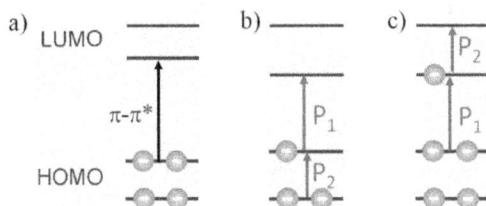

Figure 5.2: *a) In the case of a neutral homopolymer the optical bandgap is given by the π-π^* transition. When charged b) positively or c) negatively, geometric rearrangement of the polymer chain leads to new energetic levels, provoking new intra-gap optical tranisitions P_1 and P_2, which are typically located in the near- and mid-infrared spectral region.*

In the case of a neutral molecule the optical is given by the π- π^*-transition, as depicted in Figure 5.2 a) and described in section 2.1. After oxidation and the subsequent molecular reorganization towards a lower energetic minimum, the redshifted optical bandgap occurs as polaron absorption band P_1. In addition, a further polaron absorption band P_2 evolves due to a new accessible transition from a lower energetic occupied molecular orbital to the highest singly occupied molecular orbital. These transitions are typically in the range of 1 – 1.4 eV and 0.3-0.4 eV, respectively, for homopolymers used in photovoltaic applications, such as P3HT [13,98,192]. These transition energies correspond to wavelengths of ca. 1 µm and 3 µm, which are accessible by infrared (pump-probe) spectroscopy.

Figure 5.3 shows the differential absorption (ΔOD) spectrum of oxidized regioregular P3HT in solution [44]. It clearly exhibits reduced absorption up to a wavelength of 600 nm due to ground state bleaching (GB) and two absorption bands, P_1 and P_2 in the IR, proving the presence of hole-polarons. Experimental studies on chemically generated electron-polarons on P3HT are up to date not available, which is probably due to poor stability of this material upon reduction. Nevertheless, successful reduction experiments have been performed on other homopolymers like poly(phenylenephenylenevinylene) (PPPV) [91] and recently also on oligomers of fluorene[171] and oligo(phenylenevinylene) [100].

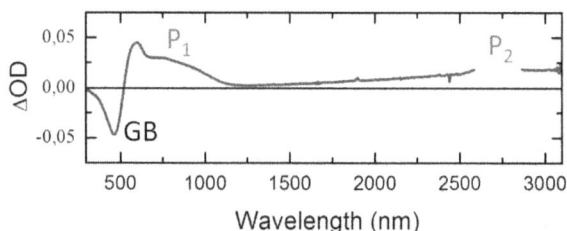

Figure 5.3: *Differential absorption (ΔOD) spectrum of regioregular P3HT solved in dichlorobenzene. Hole-polarons (cations) were generated by chemical oxidation with SbCl₅. Ground state bleaching (GB) and polaron absorption bands P_1 and P_2 are clearly visible.*

All these experiments emphasize that the absorption of cations and anions in these systems is very similar in terms of enegetic position and spectral shape. A possible explanation for this symmetry might be the symmetry of the involved molecule orbitals [100,193,194]. In the studied systems, i.e. homopolymers and oligomers, both HOMO and LUMO are mainly unstructured orbitals, which are delocalized over large parts of the chromophore. Consequently, a missing electron in the HOMO as well as an excess electron in the LUMO leads to a similar wave function distribution with similar properties. It is a valid argument that this symmetry might also hold true for other homopolymers, like P3HT. Nevertheless, in low-bandgap copolymers, with an alternating donor-acceptor structure along the molecule backbone, this symmetry is clearly broken [45]. As a consequence, symmetric absorption spectra between positive and negative polarons cannot be expected. This motivates the detailed discussion of their polaron absorption spectra in the next section, since this knowledge is crucial for a correct qualitative and quantitative interpretation of spectroscopic measurements.

5.2.2. Asymmetric polaron signatures in donor-acceptor copolymers

While it is commonly assumed, that the polaronic absorption bands can be assigned to interband (P_1) and intraband (P_2) transitions for homopolymers, as discussed in the previous section, the nature of the absorption bands in donor-acceptor copolymers is somewhat more complex, due to considerable electronic configuration mixing. Different effects originating in the sophisticated nature of such materials require a more extended model, than it is presented in Figure 5.2. Firstly, in many cases the formation of absorption bands, as they are measured experimentally, does not occur due to only one transition, but might be a superposition of several distinct transitions with different oscillator strengths. Not only different transitions but also a mixture of different transition types, i.e. inter- and intra-band, is found to contribute to this absorption band. Several higher LUMO and lower HOMO levels are found to contribute to the inter-band transition in such molecules. A superposition of different transitions, forming the absorption bands, can even be observed in homopolymers, such as poly(cyclopentadithiophene) (PCPDT) [45], with a much simpler chemical structure. However, their symmetry of electronic states above and below the optical bandgap are both of delocalized nature. This maintains the electron-hole symmetry in the distribution of the wave function and thus a qualitatively very similar absorption behavior of cations and anions. Figure 5.4 illustrates measured and calculated absorption spectra of neutral and ionized molecules of the homopolymer PCPDT and the copolymer PCPDTBT. For both materials, i.e. PCPDT in panel a) and PCPDT in panel b), the absorption of neutral molecules (red line) coincides with the spectral position of the ground state bleaching in the chemically induced differential absorption ΔOD (green line).

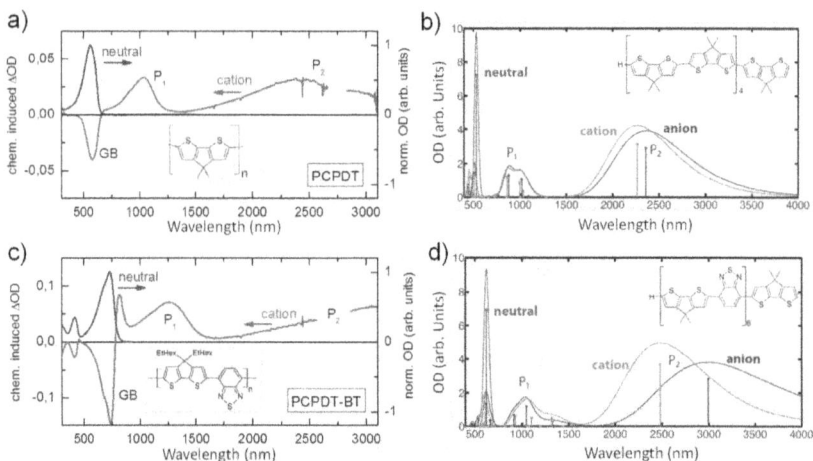

Figure 5.4: *Experimentally measured absorption of neutral (red) molecules in solution of a) the homopolymer PCPDT (solved in toluene) and c) the copolymer PCPDTBT (solved in ODCB). The related measured differential absorption upon oxidation (formed hole polarons, shown in green) shows prominent polaron absorption bands P₁ and P₂. A comparison with theoretical simulations states good agreement of measured and predicted cation spectra (green) for b) PCPDT and d) PCPDT-BT. Further comparison with calculated anion spectra (blue) shows only small differences between cations and anions for the*

homopolymer in b) but clearly broken electron-hole symmetry for the copolymer in d). Chemical structures of the measured/simulated molecules can be found as inset in the respective panel.

The latter has been realized by doping thin solutions of PCPDT and PCPDT-BT in ortho-dichlorobenzene (ODCB) and toluene, respectively, with $SbCl_5$. The subsequent ionization allowed to measure the differential absorption of the formation of cations. Prominent signatures, proofing the presence of cations, are the polaron absorption bands P_1 and P_2, which are located around 1000 nm and 2500 nm in PCPDT and around 1200 nm and 3000 nm in the case of PCPDTBT. Good agreement of the measured and calculated spectra (shown in Figure 5.4 b) and d)) of the same materials was found, although for the simulation shorter chains have been chosen for the sake of limiting computation effort.

These calculations are based on TDDFT and are published in reference [45]. Absorption spectra of neutral and positively ionized molecules could be reconstructed in a very plausible fashion, although a slight blue shift of the theoretical compared to the measured spectra remains, which is probably due to the limited chain length the used theoretical method. Chemical structures of the measured and simulated molecules can be found as insets in the respective panels. In addition to the cation absorption spectra, also the related anion absorption spectra could be retrieved by theoretical simulations. For the homopolymer in panel b), only a slight deviation is observed, proofing expected symmetry to a large extent of the involved electronic states below and above the optical bandgap, which are of delocalized nature in both cases. This symmetry is not maintained in the case of the copolymer PCPDTBT. A detailed discussion on the basis of quantum chemical calculations and experimental measurements is given in the following.

This will be done by choosing a more generalized approach using quantum chemical calculations helping to visualize all relevant electronic transitions in order to obtain a realistic view of the processes leading to polaronic absorption.

A realistic description requires to consider geometric changes of the molecule toward a new energetic minimum upon addition or removal of a charge to the molecular system, as explained in section 2.4.4. The spin-doublet structure of polarons has to be considered to influence the optical response due to their selection rules. These aspects have been caccounted for in the quantum chemical calculations carried out by Wiebeler et al., which are described in more detail in reference [45]. They are based on the introduction of natural transition orbitals (NTOs) [195], allowing us to investigate electronic excitations in the correlated many-particle system with still relatively simple terms. In principle, each transition is split up into a minimum number of pairs of effective single-particle orbitals. In ideal case, the transition can be reduced to only one relevant pair of NTOs. For more complex transitions with electron correlation playing a major role, a larger number of NTOs has to be taken into account.

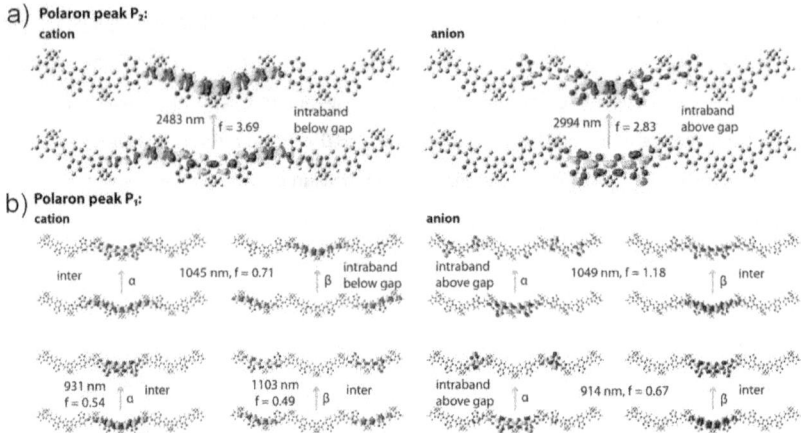

a) Polaron peak P_2:
cation — anion

2483 nm f = 3.69   intraband below gap

2994 nm f = 2.83   intraband above gap

b) Polaron peak P_1:
cation — anion

inter α 1045 nm, f = 0.71 β intraband below gap

intraband above gap α 1049 nm, f = 1.18 β inter

931 nm f = 0.54 α inter   1103 nm f = 0.49 β inter

intraband above gap α 914 nm, f = 0.67 β inter

Figure 5.5: *Natural transition orbitals (NTOs) contributing to polaron absorption bands of cations and anions of PCPDT-BT. The visualizations contain the wavelength of occurrence, the oscillator strength f, the spin sub-system (α,β), and their intra- or inter-band character. a) P_2 absorption band for cations and anions are based on one intra-band NTO pair each, being of delocalized/localized nature for cations/anions. b) P_1 shows a more complex structure, since several NTO pairs contribute to each polarity. Inter- and intra-band transitions as well as both spin sub-systems contribute to P_1. Intra-band transitions have a remarkably high charge-transfer character.*

The numerically calculated NTOs for the dominant transitions forming P_2 in the copolymer PCPDT-BT are shown in Figure 5.5 a). Wavelength and oscillator strengths f are noted in the figure as well as the different spin subsystems α and β of the involved polaronic doublet states. Both the cationic and the anionic P_2 transition are each dominated by only one single NTO pair, being located at 2483 nm and 2994 nm with respective oscillator strengths of 3.69 and 2.83. In this copolymer, states above the gap are well localized on the BT units [42] whereas states below the gap show extended π-orbital delocalized over several repeating units nature similar to those in homopolymers. This difference can be attributed to the higher electron affinity of the BT units and consequently to the broken symmetry between electronic states below and above the band gap (holes and electrons). Accordingly, the anion intra-band absorption of this copolymer mainly originates in transitions between states, which are localized on several *BT* units, whereas the cationic contribution to the P_2 peak is dominated by transitions between delocalized states. With this knowledge it is easy to understand, why the cation absorption band of PCPDT and PCPDTBT are very similar in shape and energy, while their anionic counter parts differ significantly from each other.

A slightly more complicated explanation has to be given for the higher energetic absorption band P_1 of PCPDT-BT. Nevertheless, since this absorption band is often used as a spectroscopic signature in ultrafast spectroscopic measurement with broad-band detection, it is worth to spend some time and effort to understand the slightly sophisticated nature of this absorption band. As can be seen in Figure 5.5 b), both the cationic and the anionic contribution to P_1 consist in several dominant NTO pairs. For the cation, the dominant contributions are located at 1045 nm, 1103 nm and 931 nm, with oscillator strengths of 0.71, 0.49 and 0.54, respectively. In the case of the anion, the dominant contributions are at 1049 nm and 914 nm with oscillator strengths of 1.18 and 0.67. In both cases relevant inter- and also intra-band transitions can be found, as well as NTOs belonging to the different spin sub-systems α and β. Considering the large number of transitions and much more the different nature of these transitions forming the anionic and cationic contribution to the P_1 band, the found similarity of their spectral shape is very surprising and should not be generalized to other materials of this class. Only partially, this similarity could be explained by the energy difference between HOMO and LUMO, which is independent of the character of the states below and above the gap [45,196]. Interestingly, the contributing intra-band transitions to both P_1 types show a very pronounced charge-transfer character, which means that the spatial overlap of both orbitals is strongly reduced. Since this is not observed in homopolymers, it is a first interesting indication, that charge-transfer and reduced spatial overlap of excited states plays an important role in donor-acceptor materials. Nevertheless, the large number of different contributions to the spectral shape and energetic position does not allow generalizing the gained knowledge to all donor-acceptor copolymers. For this reason, the polaron absorption of further copolymers studied in this work shall be investigated in detail in the next section.

5.2.3. Structural influences governing anion and cation absorption asymmetry

In order to get deeper insight into the properties and the behavior of polarons and polaron pairs in these novel materials for organic photovoltaics, the studies have been extended to a representative class of different donor-acceptor copolymers [44]. Figure 5.6 shows the absorption spectra of respective thin films and the chemical structure of PCPDT-BT and three further low-bandgap polymers, which have been investigated in a systematic study and will be introduced in the following.

Among the studied choice of materials is the low-bandgap copolymer PCPDT-BT, which has been introduced already in the previous sections. For sake of completeness, its neutral absorption spectrum and its chemical structure are shown again in Figure 5.6 b) and f). The electron donating CPDT unit, which is the same for all chosen copolymers, is highlighted by a blue box. The electron accepting BT unit, which is directly attached to it, is marked by a yellow circle, indicating an electron accepting unit of medium strength, more strictly spoken with moderate electron affinity. To ensure good comparability, different donor-acceptor copolymers have been chosen, sharing an identical donor unit, but varying in their acceptor moiety and/or the spatial separation of donor and acceptor centers. One further studied copolymer is poly(4,4'-bis-(2-ethylhexyl)-4H-cyclopenta[2,1-b;3,4-b']-dithiophene benzo-[1,2-c;4,5-c']bis[1,2,5]thiadiazole]) (PCPDT-BDT), as shown in   Figure 5.6 e). Its red highlighted BDT unit indicates a very electronegative acceptor. Together with PCPDT-BT it forms one category of studied materials, where the acceptor is directly linked to the donor segment. The second category consists in two copolymers, which exhibit a donor segment, being extended by two attached thiophene rings and hence further separates the centers of donor and acceptor. This category consists firstly in poly(4,4'-bis-(2-ethylhexyl)-4H-cyclopenta[2,1-b;3,4-b']-dithiophene-4,7-bis(2-thienyl)-2,1,3-benzothiadiazole)   (PCPDT-2TBT) (Figure 5.6 h)), carrying the same acceptor unit as the well known PCPDT-BT and poly(4,4'-bis-(2-ethylhexyl)-4H-cyclopenta[2,1-b;3,4-b']-dithiophene- 2,3-diphenyl-5,7-bis(2-thienyl)thieno[3,4-b]pyrazine) (PCPDT-2TTP), again with a stronger electron acceptor, as shown in Figure 5.6 g). In this second category, the CPDT donor is linked to the acceptor via thiophene spacers. We refer to them as spacers in the sense that they further separate the donor and acceptor centres of mass and are chemically similar to the CPDT donor.

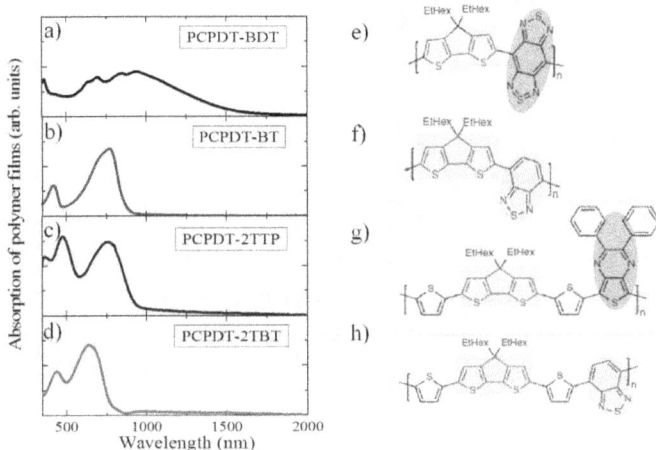

Figure 5.6: *Absorption spectra of thin films of the neutral low-bandgap copolymers a) PCPDT-BDT, b) PCPDT-BT, c) PCPDT-2TTP, d) PCPDT-2TBT. The respective chemical structures are shown in panels e) – h). All studied copolymers share the same CPDT donor unit, highlighted with a blue box, but differ in their acceptor moieties or on-chain topology. Yellow (red) indicates moderate (strong) electronegative acceptor moieties. In category one, i.e. e)-f), donor and acceptor are directly linked, while in the second category g)-h) they are further separated by thiophene rings as spacers.*

Figure 5.6 a) – d) shows the absorption spectra of the different compounds deposited as thin films. All of them exhibit an absorption onset in the NIR region of the spectrum (< 730 nm, < 1.7 eV), which is redshifted compared to homopolymers, such as P3HT. Their different acceptor moieties and chain topologies provides them different absorption spectra when compared to each other. As for other very similar copolymers [197], the energetic position of the first absorption band (S_1) is mainly determined by the $HOMO_{Donor}$–$LUMO_{Acceptor}$ energy difference and the electronic coupling between them, as discussed in more detail in section 3.3. PCPDT-BDT having the most electronegative acceptor moiety among the materials presented here, shows an absorption onset starting at about 1550 nm (~0.8 eV), as depicted in Figure 5.6 a). PCPDT-BT and PCPDT-2TBT share the same acceptor, but the presence of the thiophene spacers in PCPDT-2TBT, further separates the mass centers of the donating and accepting moieties, likely influencing the electronic structure of the coupled system. This results in a slightly blue-shifted absorption onset of PCPDT-2TBT (Figure 5.6 d)) with respect to PCPDT-BT (Figure 5.6 b)). In PCPDT-2TTP (Figure 5.6 c)), the two thiophene spacers decrease the coupling, but the stronger acceptor characteristics of TP with respect to BT appear in an S_0-S_1 absorption band located at smaller photon energy with respect to the other copolymer with spacers, PCPDT-2TBT. Table 5.1 shows measured values for the energy levels of HOMO and LUMO of the respective materials, measured by ultraviolet photoelectron spectroscopy [198]. It demonstrates the influence of electronic coupling and electronegativity of the acceptor upon the energy levels of the LUMO and the resulting optical bandgap. Even the energetic position of the HOMO is influenced by electronic coupling, although it is expected to be mainly determined by the HOMO level of the donor unit [197], which is identical for all four copolymers.

	PCPDT-BDT	PCPDT-BT	PCPDT-2TTP	PCPDT-2TBT
E_{LUMO} (eV)	-3.89	-3.57	-3.7	-3.17
E_{HOMO} (eV)	-4.89	-5.3	-5.3	-5.15

Table 5.1: *LUMO (E_{LUMO}) and HOMO (E_{HOMO}) energy values versus the vacuum energy level for the studied materials, measured by ultraviolet photoelectron spectroscopy.*

It is not devious to assume that a varying chemical structure, governing the neutral state absorption, as shown in Figure 5.6, also has a certain influence on the absorption spectra of the charged species of different donor-acceptor copolymers. For a closer investigation the absorption spectra of cations and anions for the four studied polymers have been investigated in more detail. The spectral position and shape of the hole polaron absorption bands in the infrared have been obtained for all copolymers by chemical doping with the oxidizing dopant SbCl$_5$ (compare section 4.3.1). In addition, further quantum-chemical calculations have been performed [44,99,181]. Figure 5.7 shows the corresponding dopant induced variation in the optical density, measured by successively adding small amounts of SbCl$_5$ (1 - 4% wt) to the respective polymers. As expected from the measurements shown in section 5.2.2, a recurrent feature in all spectra is the two polaron absorption bands P_1 and P_2. On the high-energy side, all the ΔOD spectra exhibit a ground-state bleaching (GB), i.e. a negative absorption change, matching the correspondent absorption profiles of the neutral molecules as commented above in Figure 5.6. For PCPDT-BT, PCPDT-2TBT and P3HT, an additional shoulder due to *EA* can be found at the onset due to electro-absorption features appearing as absorption peaks or shoulders on the blue side of the P_1 bands [96,188]. These measurements do not only reveal the spectral position of the respective polaron absorption but can further be used for a quantitative analysis of the cation absorption cross section, as will be explained in detail in the next section.

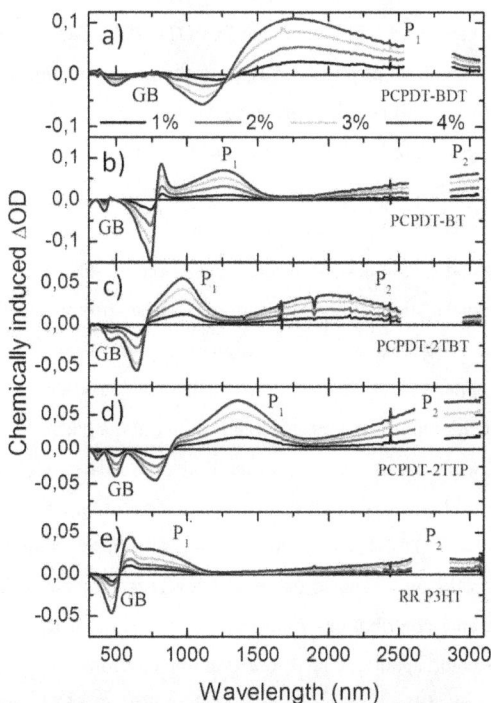

Figure 5.7: *Hole-polaron absorption spectra measured by doping the polymers with small amounts of SbCl₅ (1 – 4% wt), solved in ODCB. Chemically induced differential absorption for the donor-acceptor polymers a) PCPDT-BDT, b) PCPDT-BT, c) PCPDT-2TBT, d) PCPDT-2TTP and e) the homopolymer RR-P3HT. A linear increase of the ground state bleaching (GB) at the short wavelength side and the prominent polaron absorption bands (P₁ and P₂) in the infrared region can be observed.*

79

Because of the lacking possibility to reliably measure the absorption spectrum of anions in these materials, as detailed in sections 4.3.2 and 4.3.3, quantum chemical calculations for the absorption spectra of polarons of both polarities have been performed. To reveal the influence of the chemical structure on cation and anion absortion, a more simplfied approach [44] has been chosen. Its lower computational effort allowed to compare a larger number of different materials. For more details on this simplified theoretical approach, the reader is referred to section 4.3.4.For a direct comparison between the different polymers, we focused on chains of similar physical lengths, namely featuring four repeating units for PCPDT-BDT and PCPDT-BT and three units in the case of PCPDT-2TTP and PCPDT-2TBT. These oligomers seem to be long enough to mimic the photo-physical behavior of the corresponding copolymers. This length restriction leads to a significant blue-shift between experimentally and theoretically obtained spectra and is expected to be similar for each of the studied materials. Hence it will not be regarded as critical for this specific comparison. The resulting cation and anion spectra for the four studied oligomers are presented in Figure 5.8. Their unique combination of acceptor electron affinity and spatial donor-acceptor separation are schematically indicated as insets in the respective panels. Availability of retrieved polaron spectra for PCPDT-BT with the Hartree-Fock method as well as with the previously explained more detailed TD-DFT calculations allows a comparison of the results of both techniques and to judge on their reliability. It is worth mentioning, that both calculations show qualitatively the same result and good agreement between the retrieved spectra. Both methods predict similar spectral shapes for the P_1 band and a red-shifted P_2 absorption for the anion. It could be well reproduced, that for the PCPDT-BT cation the P_2 band originates mainly in intra-band electronic transitions from doubly occupied below-gap levels to the polaronic level, and the P_1 band involves comparable contributions from intra- and inter-band transitions. A similar distribution of transitions could again be found for the anion, confirming one major intra-band transition for P_2 and several different transitions contributing to P_1. For very low bandgap copolymers such as PCPDT-BDT as shown in Figure 5.8 a), the P_2 band of the positive polaron mainly involves electronic transitions from the polaronic level to the unoccupied levels while transitions from the occupied levels to the polaronic level mostly contribute to P_1. The assignment of the polaronic bands is therefore swapped in comparison to PCPDT-BT, as a result of the strong electron acceptor character of BDT [44]. Indeed, for PCPDT-BDT, the HOMO-LUMO bandgap is so small that when a positive or negative excess charge is added, the resulting polaronic level gets closer in energy to the LUMO or HOMO level than to the doubly occupied or unoccupied levels, respectively. In this way, the energetic order of intra- and inter-band transitions is swapped.

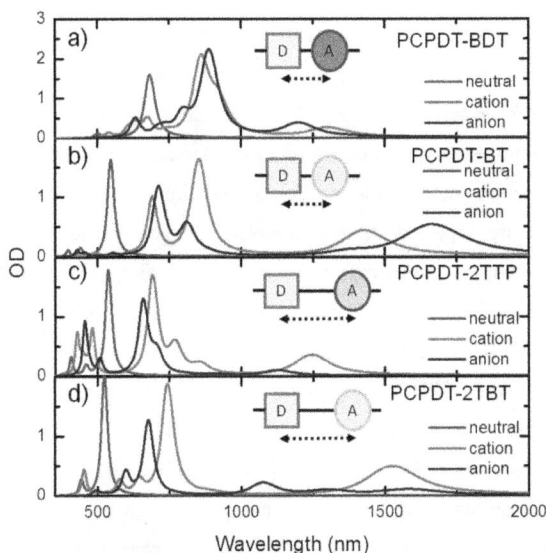

Figure 5.8: *Computed absorption spectra for neutral (red), positively charged (green) and negatively charged (blue) short oligomers sharing the chemical structure with a) PCPDT-BDT, b) PCPDT-BT, c) PCPDT-2TTP, d) PCPDT-2TBT. Electronic coupling between donor and acceptor is governed by the acceptor electronegativity (symbolized by yellow and red colors) and the donor-acceptor separation, as indicated by the inset for each material. This interaction has a large impact on the appearance of polaron absorption.*

When using thiophene spacers between the donor and acceptor units, as they are present in PCPDT-2TTP and PCPDT-2TBT in Figure 5.8 c) and d), the donor-acceptor character of the copolymer is reduced. This reduced donor-acceptor character and electronic coupling leads again to a similar behavior as PCPDT-BT, although an acceptor with a stronger electronegativity might be involved.

Overall, the optical absorption spectra computed for positive and negative polarons in the four copolymers closely resemble each other. There are, however, slight differences that can be traced back to the corresponding charge distributions. As expected, they show that the excess charge is manly confined over the donor / acceptor moieties for the cations / anions. For

PCPDT-BDT, having the strongest donor-acceptor character, the positive and negative charge is almost exclusively (> 75%) localized on three donor or acceptor units, respectively. It is straight forward to explain this as a consequence of the large difference in the ionization potential and electron affinity between the two moieties. Thus, in this case, a very similar confinement in the polaronic levels is predicted that in turn leads to very similar optical absorption spectra for excess and lacking electrons. Due to the large discrepancy in electron affinity and the very similar spatial extension of the donating CPDT and the accepting BDT unit, the wave functions of electron and hole show similar confinement, and thus similar absorption spectra are expected. Almost complete charge localization over the electron-donating (> 80%) and, to a lesser extent, electron-withdrawing (>55%) units takes place in PCPDT-2TTP and PCPDT-2TBT when merging the charge distributions computed for the PCPDT units and the neighboring thiophene rings. There, the increased delocalization of the positive charge over the PCPDT-2T segments yields increased absorption cross sections for the positive polaron (compare Table 5.3). In PCPDT-BT, the excess positive (negative) charge is mostly confined (>65%) over the PCPDT (BT) units. However, because of the strong electronic coupling compared to the energy mismatch between the frontier energy levels for this donor-acceptor combination, the positive charge partly leaks out over the neighboring BT units. A similar yet smaller effect is observed for the negative charge. As a result, an electron-hole asymmetry is also in this copolymer, although less pronounced than for the copolymers incorporating thiophene rings [44,45].

For a correct qualitative interpretation of spectroscopic measurements the previously provided information about origin and spectral shape of polaron absorption bands is sufficient. In order to extend the extraction of information of those measurements to a quantitative level, detailed knowledge of relevant material properties, i.e. the molecular absorption coefficient and the absorption cross section, for the investigated absorption band is required. Next section will focus on retrieving these important parameters for the four investigated copolymers and allow a further quantitative analysis of polaron pair formation in these systems.

5.2.4. Absorption cross section of polaron pairs

Quantitative evaluation of spectroscopic measurements allows retrieving a much higher level of information about the studied system than pure qualitative evaluation. However, for achieving measurement data, providing reliable quantitative information, a high level of experimental accuracy and a meticulous record of all device and system parameters are required. In addition, material parameters, like molar extinction coefficient or the absorption coefficient, are required, which are in most cases unknown for novel material systems like most donor-acceptor copolymers.

One goal of this work is, to achieve a deep understanding of polaron pair formation in low-bandgap polymers, as discussed in detail in the following chapter. For this purpose, quantitative measurements are a precious tool for comparing the polaron pair formation efficiency of different materials and to retrieve the influencing factors governing their yield. To do so, this section will focus on the determination of the absorption cross section of the polaron absorption bands in these materials. As already discussed in the previous section, the successive doping of polymers in solution with a small and controlled amount of a strong oxidant leads to stepwise ionization of the polymer. This is going along with the occurrence of their respective polaron absorption bands in the differential absorption spectrum, as shown in Figure 5.7. For details about the experimental procedure the reader is referred to section 4.3.1. All four studied copolymers and the homopolymers P3HT show a linear increase of their spectral features occurring with ionization, up to a doping ratio of 4% of the dopant with respect to the weight of the polymer solved in the solution. This linearity is maintained for all observed spectral features, i.e. ground state bleaching (GB) and the polaron absorption bands P_1 and P_2. Above a doping ratio of 5% the signal increase drops below linearity which indicates a set in of saturation effects. Staying in the linear regime and assuming that one cation is formed for each added dopant molecule, allows to calculate the absorption cross section of each position of the differential spectrum, including the polaron absorption bands. To do so, we use Lambert-Beer's law,

$$I = I_0 e^{-\alpha x} \tag{5.1}$$

describing the transmitted light intensity I, which remains from the initial intensity I_0 after suffering from absorption losses during the propagation through the sample. Hereby, α denotes the absorption coefficient of the transmitting object and x its thickness. This is related to the material specific molar absorption coefficient ε by

$$\log_{10}\left(\frac{I}{I_0}\right) = -\alpha x = \varepsilon(\lambda_{Probe}) \cdot C_{Dopant} \cdot x = OD_{Polaron}(\lambda_{Probe}) \tag{5.2}$$

In detail, it relates the measured optical density of the polaron absorption signal $OD_{Polaron}$ at a suitable detection wavelength λ_{Probe} with the material parameter we are interested in, i.e. the molar absorption coefficient ε. Thus, the latter can be determined by measuring the optical density for a known concentration of polarons C_{Dopant} (given by the controlled doping concentration) and the known sample thickness x. The macroscopic material property ε is given by

$$\varepsilon(\lambda_{Probe}) = \frac{OD(\lambda_{Probe})}{C_{Dopant} \cdot x} \tag{5.3}$$

which relates to the microscopic polaron (cation) absorption cross section $\sigma_{Polaron}$ as depicted in Equation (5.4).

$$\sigma_{Polaron}(\lambda_{Probe}) = ln(10) \cdot \frac{\varepsilon(\lambda_{Probe})}{N_A} = ln(10) \cdot \frac{OD(\lambda_{Probe})}{C_{Dopant} \cdot x \cdot N_A} \tag{5.4}$$

Consequently the cation absorption cross section depends on the fraction of the measured differential absorption signal at a chosen wavelength to the related dopant concentration. This fraction can be retrieved for each polymer from the doping measurements, shown in Figure 5.3. Furthermore, it depends on the sample thickness x and the Avogadro constant N_A.

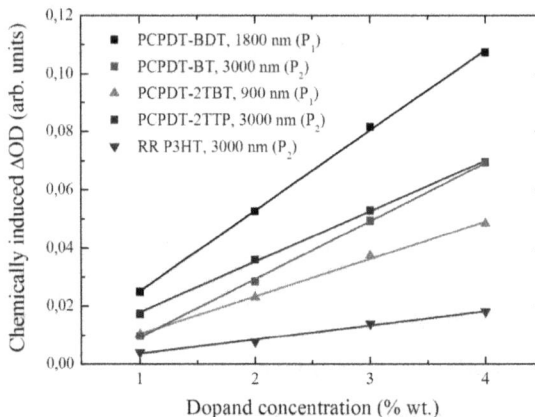

Figure 5.9: *Chemically induced ΔOD value of the chosen polaron absorption band for all studied copolymers and P3HT, versus the concentration of added dopant. Each measurement point corresponds to the signal amplitude at the respective wavelength in the curves shown in Figure 5.7. The wavelengths have been chosen close to the maximum of the polaron band, which is accessible to an ultrafast spectroscopic investigation.*

Figure 5.9 shows a plot of the ΔOD amplitude of cation absorption, taken from the curves plotted in Figure 5.3, versus the related doping concentration for each polymer. Each percent of dopant corresponds to a concentration of $5.02 \cdot 10^{-10} \frac{\text{mol}}{\text{cm}^3}$ for the chosen concentration of polymer solution, as explained in detail in section 4.3.1. The wavelength for determining the absorption cross section of each polymer has been chosen close to the maximum of the respective polaron absorption band, which will be used as a suitable probe wavelength for the detection of polaron pairs with ultrafast femtosecond spectroscopic methods.

The fraction of ΔOD versus the doping concentration can be retrieved as the inclination of the straight lines in Figure 5.9. Furthermore, this plot is a clear proof that the conducted experiments with doping ratios up to 4% are still in the linear regime and no saturation effects are present. Table 5.2 summarizes the retrieved data on the absorptivity of cations of the respective copolymers and P3HT.

Polymer	λ_{Probe} (μm)	ε (cm^2/mol)	σ_{Cation} (cm^2)
PCPDT-BDT	1.8	$(5.5 \pm 0.1) \cdot 10^7$	$(2.11 \pm 0.04) \cdot 10^{-16}$
PCPDT-BT	3	$(4.0 \pm 0.1) \cdot 10^7$	$(1.52 \pm 0.02) \cdot 10^{-16}$
PCPDT-2TTP	3	$(3.5 \pm 0.1) \cdot 10^7$	$(1.36 \pm 0.03) \cdot 10^{-16}$
PCPDT-2TBT	0.9	$(2.6 \pm 0.2) \cdot 10^7$	$(9.8 \pm 0.6) \cdot 10^{-17}$
RR- P3HT	3	$(9.7 \pm 0.7) \cdot 10^6$	$(3.7 \pm 0.3) \cdot 10^{-17}$

Table 5.2: *Summary of retrieved data on the polaron absorptivity of all four studied donor-acceptor copolymers and P3HT. The molar absorption coefficient ε and the absorption cross section for the cations, σ_{Cation}, at the wavelength of interest λ_{Probe} are shown.*

Absorption cross section σ_{Cation} and the molar absorption coefficient ε have been retrieved by doping measurements and calculations as discussed above. For the cation absorption cross section of P3HT several values can be found in literature [13,199–202], ranging from $5 \cdot 10^{-17}$ to $1 \cdot 10^{-15}$. The wide spread estimates clearly demonstrate the difficulty of measuring these values in a reliable fashion. All of these values, which can be found in literature so far, aimed for the absorption cross section of P_1 for P3HT in the NIR, for instance at 850 nm as mentioned in reference [200]. The experiments in this work aim for measurements further in the IR spectral region, the absorption cross section has been evaluated at appropriate wavelengths, as shown in Table 5.2. Regarding the weaker oscillator strength of P_2 compared to P_1, the lower measured value for the absorption cross section than those found in literature is not unexpected and is regarded as a good indication for the reliability of the performed measurements.

Together with the theoretical comparison of cation and anion absorption spectra, which is discussed in section 5.2.3, an estimate of the absorption cross section of the related anion is possible and given in the following.

Table 5.3 depticts the intensites of polaron bands obtained by the combined oscillator strengths for transitions located in a close vicinity (≈ 0.2 eV) to P_1 or P_2.

Polymer	Polaron band	I_{Cation}	I_{Anion}
PCPDT-BDT	P_1	3.05	3.32
PCPDT-BT	P_2	0.47	0.65
PCPDT-2TTP	P_2	0.38	0.1
PCPDT-2TBT	P_1	2.63	1.8
RR- P3HT	P_2	1	1

Table 5.3: *Relative absorption strengths of cations and anions. The values have been obtained from the calculated polaron absorption spectra shown in Figure 5.8. The relative intensities of polaron bands are retrieved by the combined oscillator strengths for transitions in close vicinity to the related band P_1 or P_2. For the reasons commented above, the polaron absorption of the homopolymers P3HT is supposed to exhibit electron-hole symmetry.*

The fraction of these values gives a scaling factor for the absorption strength of the anion compared to the cation for each polymer and consequently allows obtaining the absorption cross section for anions and polaron pairs as depicted in equations (5.5) and (5.6) [44].

$$\sigma_{Anion}(\lambda_{Probe}) = \frac{I_{Anion}}{I_{Cation}} \cdot \sigma_{Cation}(\lambda_{Probe}) \tag{5.5}$$

$$\sigma_{PP}(\lambda_{Probe}) = \sigma_{Cation}(\lambda_{Probe}) + \sigma_{Anion}(\lambda_{Probe})$$

$$= \left(1 + \frac{I_{Anion}}{I_{Cation}}\right) \cdot \sigma_{Cation}(\lambda_{Probe}). \tag{5.6}$$

As explained in section 2.4.4, the absorption cross section of polaron pairs can be regarded as the sum of their constituents, because of negligible interaction between them [47].

Altogether, this experimental and theoretical approach allows to estimate the absorption cross section of cations, anions and polaron pairs. This precious information will help retrieving quantitative information from spectroscopic measurements. Using this benefit, the following chapters will focus on the formation yield of polaron pairs upon photoexcitation of low-bandgap polymers and investigate the influences of their chemical structure and the role of excess excitation energy.

5.3. Chapter summary

In Chapter 5 we investigated the spectroscopic signatures of excited states in organic materials, whose detailed knowledge is of utmost importance for the evaluation and interpretation of spectroscopic measurements on an ultrafast timescale. The formation and decay of excitons can be observed by measuring the stimulated emission (SE), occurring via radiative decay of an exciton from excited state S_1 to the ground state S_0. In addition, the presence of an excitonic absorption band (EX) proves their existence, which occurs due to further excitation from S_1 to S_n (n = 2, 3,...) by absorption of a photon. These excitonic features are similar in their nature for many homo- and copolymers, although their energetic position might be shifted according to the position of the ground state energy levels.

Polarons and polaron pairs show unique spectral absorption bands P_1 and P_2, which are located in the NIR and MIR spectral region. For homopolymers, P_1 and P_2 originate in an inter- and intra-band transition, respectively, which are very similar for excess positive (cations) and negative (anions) charges, due to symmetry of the contributing orbitals above and below the bandgap (electron and hole). By using different theoretical methods, we found this symmetry to be broken in donor acceptor copolymers because of the different orbitals being involved for charges of different sign. In this case, for both polaron absorption bands, intra- and inter-band transitions are involved, requiring a more sophisticated interpretation than homopolymers. A large number of various contributing orbitals breaks the electron hole symmetry and leads to absorption bands which occur in different shape and strength for cations and anions. Structural correlations of the copolymers exhibit a strong influence on the wave function overlap, governing the broken symmetry for polarons of different polarity. The calculated absorption spectra of cations and anions of four studied copolymers PCPDT-BDT, PCPDT-BT, PCPDT-2TTP and PCPDT-2TBT together with their experimentally retrieved cation absorptions, allowed to retrieve values for the polaron pair absorption cross section. This knowledge permits a detailed quantitative evaluation of the polaron formation yield, measured in ultrafast spectroscopic measurements, what will be discussed in the following chapter.

6. Enhanced polaron pair formation in donor-acceptor copolymers

The unique optical and electronic properties of organic semiconductors often originate from a tailored chemical structure. This allows directed optimization of conjugated macromolecules towards desired properties for organic electronics. As discussed in section 3.3, donor-acceptor copolymers are the result of an optimization towards an enhanced absorption in the red and NIR spectral region. The low bandgap offers optimum harvesting of sunlight and has inspired work towards record power conversion efficiencies [123,155]. It is straight forward to assume, that the alternating donor-acceptor structure along the polymer backbone not only affects light absorption, but also has a large impact on the formation and characteristics of photoexcitations. In this chapter, the formation yield of weakly bound polaron pairs is described with special focus on the influences of the material's chemical structure.

6.1. Intrinsic polaron pair formation in pristine polymer films

Although the power conversion efficiency of organic solar cells has been demonstrated to depend on a number of different factors, such as morphology of the blend-film [203], carrier mobility [204], and device architecture [205], the first fundamental processes remain photon absorption by a chromophore and electron-hole pair dissociation, as described in detail in section 2.4. While light absorption has been largely improved towards the near infrared (NIR) spectral region by the introduction of low-bandgap copolymers [18,19], its influence on exciton dissociation remains largely unexplored. So far, the charge dissociation process has been mainly investigated in homopolymers [6,10–12]. Those results cannot be fully applied to copolymers due to their very different chemical structure.

The basic model considers light absorption to generate strongly bound Frenkel excitons (binding energy > 0.5 eV), which then dissociate primarily at the interface with the fullerene in photovoltaic blends [128,206]. There, the large energy offset ($\geq$ 0.5 eV) between the LUMOs of polymer and fullerene (compare Figure 3.5) drives exciton dissociation and charge separation. This might possibly be gated by the formation of hot delocalized states [12]. Thus, charge dissociation is supposed to occur only after diffusion of the excitons in the polymer. However, several experimental and theoretical studies suggested that the primary photoexcitations in pristine conjugated polymers are not only excitons, but also polaron-pairs [16,17,207]. The latter ones have been reported to play a significant role in photovoltaic action and are envisaged to be as important as excitons, provided their yield is substantial [200,208,209]. These and other optical investigations were focused on the dynamics of excitons [210] or charge separated states when the polymer is mixed with the fullerene [211–214]. The question, of what the polaron pair generation-yield in pristine donor-acceptor copolymers actually is, remains largely unanswered, while being strongly relevant for the understanding and improvement of these materials toward their application in solar cells. For example, the previously disregarded contribution of polaron pairs, forming directly in the polymer may lead to the discrepancies between theoretically predicted solar cell efficiencies, based on Onsager-Braun models, and the higher measured values [10,204].

Also in homopolymers like P3HT, a small percentage of the formed photoexcitations results in polaron pairs. In the case of homopolymers, the branching ratio of charged polaron pairs to neutral excitons depends significantly on energetic disorder in the system. This can be either due to the presence of different aggregates of polymer chains, providing energetic disorder between microcrystalline structures [13,188], or directly due to an interface between amorphous and crystalline domains [57]. Nevertheless, the values for the branching ratio in P3HT found in literature vary strongly [13,188,207,215], from < 1% up to 30%. This does not allow to draw a consistent picture of the charge carrier formation mechanism and the influencing factors. This confusion is further promoted by the vast range of values for the absorption cross sections of polarons found in literature, as already discussed in section 5.2.4.

The combination of different experimental and theoretical efforts allows to reliably measure the polaron pair formation yield. In this work we present the results for several different copolymers together with quantitative numbers. It will be demonstrated in the following sections, in how far the inherent energetic disorder, provoked by the alternating structure of donor and acceptor moieties along the backbone of copolymers, has an influence on the formation of polaron pairs. This knowledge might be helpful for a precise optimization of such materials towards higher polaron pair yields. As a consequence, it might result in an increase of photovoltaic efficiency when extracting polaron pairs instead of the remaining strongly bound excitons.

6.2. Exciton to polaron pair branching ratio

Since the goal of any photovoltaic device is the extraction of separated charge carriers, their formation upon light absorption remains one of the most interesting and relevant topics to the photovoltaic research community. Due to the novelty of donor-acceptor materials, the photophysical behavior of donor-acceptor copolymers is still mainly unexplored. In this section we discuss the formation yield of weakly bound polaron pairs in pristine thin films of donor-acceptor copolymers without addition of any electron accepting material. As an excellent tool for studies of this kind, a femtosecond pump-probe setup has been used, providing ultrashort excitation pulses at 800 nm or 400 nm and probe pulses tunable in the NIR and MIR. This allows to address the polaron absorption bands P_1 and P_2. While for P_1 a spectral overlap with other spectral signatures, such as stimulated emission, is possible, the P_2 absorption band allows a direct detection of polaronic excitations without interference of any other species. Figure 6.1 shows the retrieved optically induced differential absorption ΔOD (blue dots) at zero-delay after optical excitation at 800 nm for PCPDT-BDT, PCPDT-BT and PCPDT-2TTP in panels a)-c) and for PCPDT-2TBT after excitation with 660 in panel d).

An overlay of the chemically induced ΔOD (solid lines) of the corresponding cations, as described in section 5.2, shows good agreement of the spectral shape of the optical and chemical measurements. It is a direct proof that upon light absorption not only strongly bound neutral excitons, but also weakly bound polaron pairs are formed. Both species are already present after 150 fs, which is the temporal resolution of the setup. More details on their formation, especially whether excitons are a preceding state prior to their formation, may not be given due to the limited temporal resolution of the pump-probe setup (ca. 150 fs, compare experimental section).

One further spectroscopic signature, being exclusively available after optical excitation, is the absorption band of neutral exctions (*EX*). Its origin is explained in section 5.1.

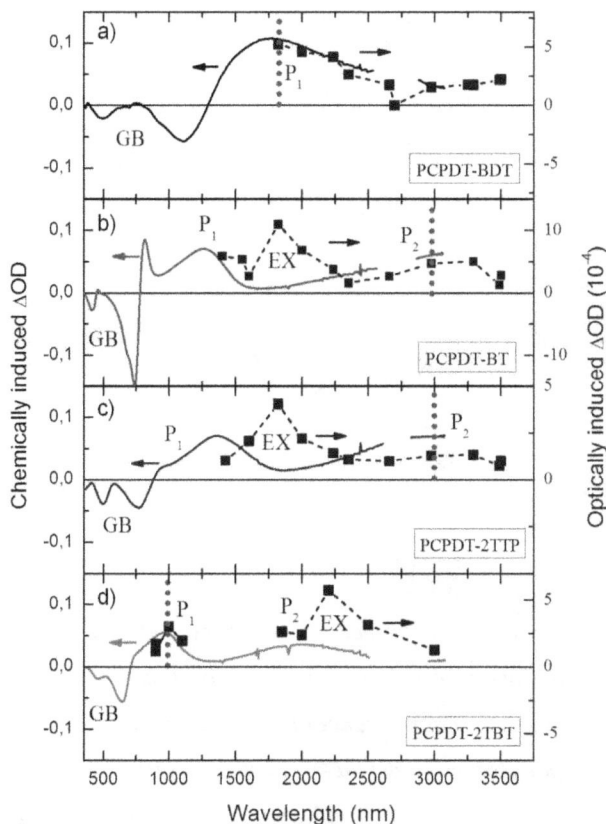

Figure 6.1: *Optically induced ΔOD (blue dots) of a) PCPDT-BDT, b) PCPDT-BT, c) PCPDT-2TTP, d) PCPDT-2TBT. The latter has been excited at a wavelength of 660 nm, all other copolymers at 800 nm. The optically induced ΔOD at zero time-delay is in good agreement with the ΔOD of chemical induced cations, proving the ultrafast formation of polaron pairs upon photoexcitation. As an additional feature in the case of optical excitation, the absorption band of neutral excitons (EX) can be seen. Vertical red dotted lines mark the wavelengths used for a further quantitative analysis.*

The differential absorption signal retrieved with optical excitation obviously leads to the formation of both excitons and polaron pairs at an ultrafast timescale with a polaron pair yield η, while upon chemical ionization no excited states are formed. A simplified physical picture of the two alternative formation and decay paths is given in Figure 6.2.

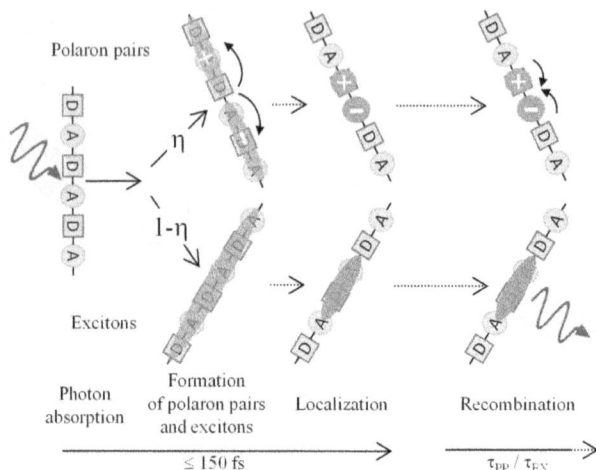

Figure 6.2: *Illustration of the formation of photoexcited states in chromophores of donor-acceptor copolymers. Upon photon absorption, within 150 fs, either a strongly bound neutral exciton (orange) is formed with a yield 1-η or a weakly bound polaron pair (green and blue) with yield η. Both species might be of delocalized nature, undergoing localization after photogeneration. Polaron pairs show recombination dynamics different from those of neutral excitons. D and A indicate the donor and acceptor segments along the copolymer backbone.*

After absorption of a photon, a polaron pair, consisting of an electron- and hole-polaron, is formed on the chain with a yield η. Their nature might be localized or delocalized. After subsequent localization of the two constituents, they recombine to the ground state with a lifetime of τ_{PP}, which is characteristic for each material. The larger part of photoexcitations, with the complementary yield of $1-\eta$, results in the formation of neutral excitons, which are delocalized over the whole chromophore. After localization, they decay with lifetime of τ_{EX}, which is typically distinct to the lifetime of polaron pairs τ_{PP}. Different lifetimes are a clear proof for the distinct character of these two species. The formation of both, excitons and polaron pairs within the temporal resolution of the setup is consistent with the picture drawn from previous studies on homopolymers [15,188]. However, a more precise statement on the nature of the primary photoexcitations, that is, whether excitons decay into polaron pairs or the latter are the primary photoexcitations, would require a time resolution well below 100 fs. For the sake of clarity, in Figure 6.3 a) - c) normalized $\Delta T/T$ spectra of PCPDT-BDT, PCPDT-BT and PCPDT-2TTP are plotted. A clear homogeneous decay of the polaron absorption is observed at several time-delay steps, without any recognizable spectral shift.

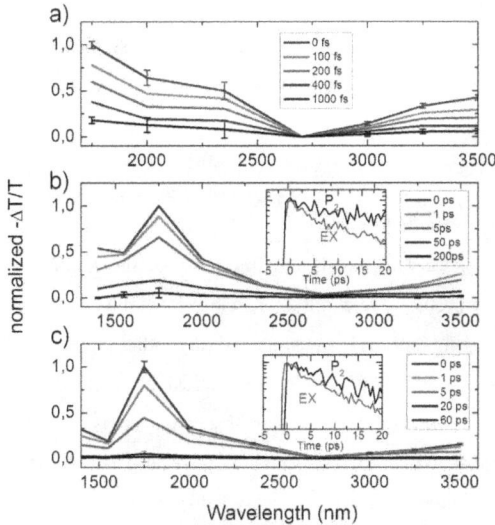

Figure 6.3: *Temporal evolution of the $\Delta T/T$ spectra for a) PCPDT-BDT, b) PCPDT-BT and c) PCPDT-2TTP. The spectrum is plotted for several discrete time-delay steps during the decay of the photoexcitations. The decay dynamics plotted in the insets in panel b) and c) demonstrate different decay dynamics of polaron pairs and excitons in the respective materials.*

Due to technical limitations this measurement could not be conducted for the copolymer PCPDT-2TBT. In the insets of Figure 6.3 b) and c), the decay dynamics of P_2 and *EX* are plotted for the respective copolymer. The polaron pairs clearly show longer decay dynamics than the excitons, proving the different nature of both species. A similar comparison for PCPDT-BDT could not be addressed, because the *EX* absorption band was not accessible in the spectral detection window of the setups.

This technique, together with the absorption cross section of the corresponding polaron bands, delivers a quantitative way to evaluate the polaron pair formation yield η of our four investigated copolymers and P3HT.

The ratio of polaron pairs among the generated photoexcitations is determined in a quantitative fashion. For this, it is of utmost importance to carefully measure the incident excitation laser pulse energy $I_{Excitation}$ and the optical densitiy of the thin polymer film OD_{Sample}. Assuming that every absorbed photon generates one photoexcitation in the polymer, allows to calculate the number of formed photoexcitations, $N_{Excitations}$, by

$$N_{Excitations} = \frac{I_{Excitation}}{\hbar\omega_{Pump}} \cdot \left(1 - 10^{\left(-OD_{Sample}(\lambda_{Pump})\right)}\right) \tag{6.1}$$

where λ_{Pump} and ω_{Pump} are wavelength and frequency of the excitation pulse, respectively. Suitable choice of the probe wavelength λ_{Probe}, close to a maximum of a polaron absorption band with no interfering spectral signature of any other species, allows to retrieve a differential absorption signal $\Delta T/T$, originating exclusively from the absorption of polaron pairs. Together with the related polaron pair absorption cross section, σ_{PP}, as it has been derived in section 5.2.4, the number of generated polaron pairs can be determined for every time delay τ. The maximum signal at zero-delay between pump and probe pulse, as it can be retrieved from Figure 6.4, leads to the number of polaron pairs formed by the excitation pulse

$$N_{PP} = -\frac{1}{\sigma_{PP}(\lambda_{Probe})} \cdot \log_{10}\left(1 - \frac{\Delta T(\tau_0, \lambda_{Probe})}{T}\right). \tag{6.2}$$

Finally, we can evaluate the polaron pair formation yield η, also called branching ratio, as:

$$\eta = \frac{N_{PP}}{N_{Excitations}} \tag{6.3}$$

Figure 6.4 shows the formation yield in %, normalized by the number of photoexcitations, and the decay of polaron pairs on an ultrafast timescale. The respective excitation wavelength has been chosen to be 800 nm for PCPDT-BDT, PCPDT-BT and PCPDT-2TBT in panels a) - c), 660 nm for PCPDT-2TBT in panel d) and 400 nm for the homopolymer P3HT, shown in panel e). An overview of the probe wavelengths together with the related absorption cross section is given in Table 5.2, and is also indicated in the respective spectrum (dotted red line in Figure 6.1, not shown for P3HT). For P3HT an excitation wavelength of 400 nm and a probe wavelength of 3000 nm have been chosen. This is a convenient wavelength for probing P_2 as obvious from Figure 5.3.

While for the homopolymer P3HT a polaron pair formation yield of 8% ($\pm$2%) at zero-delay between pump and probe pulse τ_0 was measured, for the studied copolymers significantly higher yields of 24% ($\pm$1%) for PCPDT-BDT, 20% ($\pm$1%) for PCPDT-BT, 23% ($\pm$2%) for PCPDT-2TTP and 15% ($\pm$1.5%) for PCPDT-2TBT were found. This is up to a factor of three higher than for P3HT. For comparison, for a PCPDT-2TBT:PCBM blend (1:1) a polaron yield of 48 % was measured. In a blend with PCBM only cations are present on the polymer molecules, while electrons are transferred to PCBM molecules with an efficiency of almost 100% [123]. Consequently, a maximum yield of polaron signal of 50 % is possible and a value close to it is expected. The found yield of 48 % underlines the reliability of the chosen method for the determination of the polaron pair yield. In contrast to time-averaging experimental methods, this ultrafast spectroscopic technique allows to obtain information about the maximum yield independent of the species' lifetime.

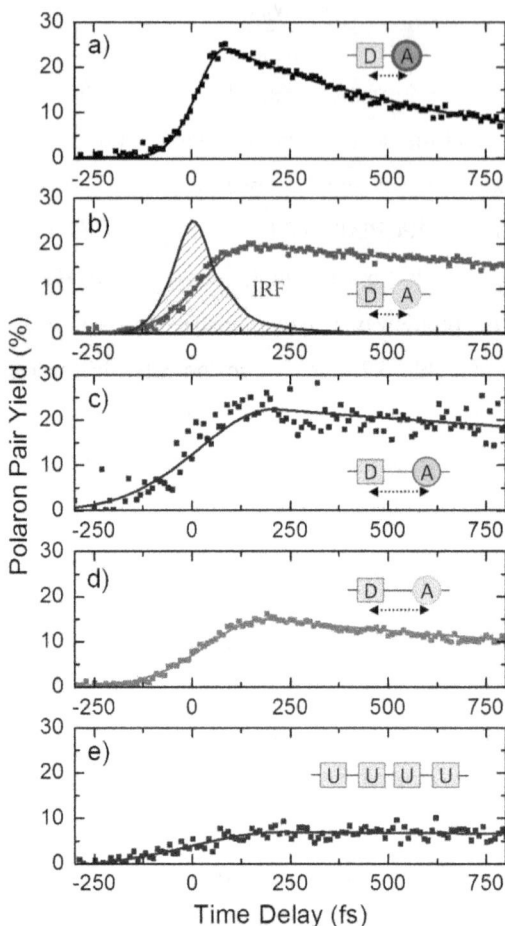

Figure 6.4: *Time traces of polaron pair generation, directly after photoexcitation, for the copolymers a) PCPDT-BDT, b) PCPDT-BT, c) PCPDT-2TTP, d) PCPDT-2TBT and the homopolymer e) P3HT. Their respective polaron pair generation yields, given in the same order, are η = 24% ($\pm1\%$), 20% ($\pm1\%$), 23% ($\pm2\%$), 15% ($\pm1.5\%$) and 8% ($\pm2\%$). The insets illustrate the variations in the polymer repeating units in terms of donor-acceptor centre of mass distance (arrows) and acceptor strength (colour coded yellow to red). The inset in b) shows the pump-probe instrument response function (IRF) for 800 nm*

pump and 3000 nm probe. Solid lines are fit to the data points which are shown as squares.

Thus it provides separate information about yield and recombination dynamics. It is worth to mention, that all polaron absorption signals rise within the limited temporal resolution of the pump-probe setup, which is ca. 150 fs (FWHM), as shown by the instrument response function (IRF) in panel b). These measurements show, that photoexcitations in donor-acceptor copolymers exhibit a higher percentage of weakly bound polaron pairs, with respect to P3HT. Due to their reduced binding energy, compared to strongly bound neutral excitons, their extraction from blends with suitable acceptor materials might result in a lower voltage loss than extraction of excitons. This could lead to higher efficiencies for polymer solar cells. Further, these measurements reveal significant differences in the polaron pair formation yield and lifetime between the four studied copolymers. An overview over these differences is given in the next section, with a strong focus on the influencing structural parameters.

6.3. Structural correlations in the generation of polaron pairs

In this section, we present a systematic study of the yield and decay dynamics of polaron pairs in the four studied donor-acceptor copolymers. Our special interest is to understand how the chemical structure influences the parameters, critical for photovoltaic. The four copolymers share the same CPDT donor units but vary in their acceptor units and/or are provided with additional thiophene spacers, as explained in detail in section 5.2.3. This is an excellent set of materials to investigate the influences of acceptor electron affinity and spatial extension on polaron pair formation. Both parameters strongly influence the electronic coupling and the formation of new energy levels (compare section 3.3). The model system P3HT has been chosen for comparability, to place the results of this investigation in a context of photovoltaics.

Ultrafast spectroscopic studies, exciting the materials close to the optical bandgap and probing a polaron absorption band at a suitable IR wavelength, allowed to retrieve the polaron pair formation yield on a femtosecond timescale, as shown in Figure 6.4, and explained in section 6.2. The decay dynamics of the polaron pair signals on a longer time scale reveal significant differences in the recombination dynamics of polaron pairs in the different copolymers. In Figure 6.5, the long time scale decay dynamics of PCPDT-BDT (black squares), PCPDT-BT (red dots), PCPDT-2TTP (blue triangles) and PCPDT-2TBT (green triangles) are plotted on a logarithmic scale, together with the best fitting bi-exponential decays curves.

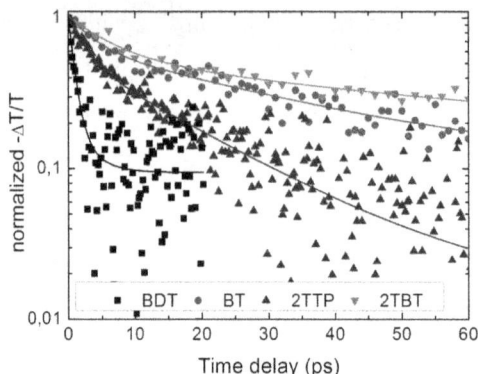

Figure 6.5: *Recombination dynamics of polaron pairs on a 60 ps timescale of a) PCPDT-BDT (black squares), b) PCPDT-BT (red dots), c) PCPDT-2TBT (blue triangles) and d) PCPDT-2TBT (green triangles). Excitation and probe wavelengths are the same as for the measurements of Figure 6.4. Solid lines of the respective color show best fitting curves for the data points, assuming a bi-exponential decay. All parameters are given in Table 6.1.*

The fitting parameters together with the polaron pair yield and the chemical structure properties are summarized in Table 6.1. Hereby, A_n and τ_n denote the amplitudes and decay times of the bi-exponential decay, respectively, and y_0 the residual long-living signal strength. The summary in Table 6.1 allows a comprehensive interpretation of the optoelectronic response after photoexcitation and a comparison between these materials with a varying chemical structure. In all cases, the rise time follows the response time of our setup at the different pump and probe wavelengths. Low-bandgap copolymers with strong electronegative acceptors, such as PCPDT-BDT and PCPDT-2TTP (Figure 6.4 a) and c)), exhibit the largest probability of polaron pair generation after photoexcitation, with 24% and 23%, respectively.

Acceptor	Struct.	η (%)	A1	τ_1 (ps)	A2	τ_2 (ps)	y_0
BDT	D-A	24 ± 1	0.71 ± 0,06	**0.4 ± 0.1**	0.19 ± 0.04	**3.4 ± 2.0**	0.07 ± 0.03
BT	D-A	20 ± 1	0.4 ± 0.05	**3.7 ± 1.0**	0.55 ± 0.04	**32.5 ± 4.7**	0.09 ± 0.02
2TTP	D---A	23 ± 2	0.45 ± 0.05	**2.5 ± 0.5**	0.54 ± 0.04	**16.7 ± 1.3**	0.15 ± 0.01
2TBT	D---A	15 ± 1.5	0.48 ± 0.06	**6.5 ± 1.4**	0.41 ± 0.04	**57 ± 18**	0.14 ± 0.04

Table 6.1: *Summary of the polaron pair formation yield η and the fitting parameters from the best fitting bi-exponential curves shown in Figure 6.5. Further, the acceptor unit and a simplified indication of the material's structural properties is given. All dynamics consist in an initial rapid decay and a longer decay. They differ for the different copolymers by more than one order of magnitude.*

In contrast, PCPDT-BT and PCPDT-2TBT with weaker acceptor units exhibit smaller yields of 20% and 15%, respectively. The influence of thiophene spacers between donor and acceptor can be directly seen when comparing PCPDT-BT and PCPDT-2TBT, because they share identical acceptor units. The spatial separation between the donor and acceptor centre of mass leads to a drop of the formation yield from 20% to 15%. Interestingly, also the dynamics of the recombination process differ and can be addressed to some extent to acceptor strength and its separation from the donor. Polymers including thiophene spacers exhibit a longer initial recombination time about 2.5 and 6.5 ps for PCPDT-2TTP and -2TBT, respectively. In contrast the copolymers without spacers show a faster recombination decay of 0.41 and 3.7 ps for PCPDT-BDT and -BT, respectively. The same trend is continued in the decay times of the slower decay components, which is only 3.4 and 32.5 ps for -BDT and -BT, respectively, but increases to 16.7 and 57 ps for -2TTP and -2TBT.

Acceptor strength has an even stronger effect on yield and lifetime of polaron pairs. We compare copolymers within each category, i.e. with and without thiophene spacers. It shows that the recombination lifetime is strongly dependent on the electron affinity of the acceptor. For instance, PCPDT-2TTP and PCPDT-BDT, which incorporate very strong acceptors, show faster recombination times compared to PCPDT-2TBT and PCPDT-BT with weaker acceptors. The more pronounced shortening of initial rapid and long decay times for stronger electron acceptors than when removing the spacers, leads to the conclusion, that the electron affinity has a higher influence on yield and lifetime than the effect of the spacers. Furthermore, the signal from polaron pairs in P3HT (Figure 6.4 e), monitored at 3000 nm equivalent to reference [216]), exhibits the lowest yield among the investigated materials. The measured yield of 8% ($\pm$2%) is comparable to values obtained with complementary techniques [217,218]. This last comparison supports the reasoning that donor and acceptor moieties within the polymer backbone are promoting charge separation and polaron pair formation [44], whereas in homopolymers, such as P3HT, charge separation is less probable and probably driven by energetic disorder between different polymer chromophores [57,58].

In summary, the obtained results demonstrate that copolymers with strong accepting units show a higher formation yield and a shorter lifetime of polaron pairs, than copolymers with moderate acceptors. In addition, the presence of thiophene spacers, which further separates the donor and acceptor centers, reduces the polaron pair yield but significantly increases their lifetime. The influence of acceptor strength is found to be larger than the effect of adding a thiophene spacer between donor and acceptor.

These results emphasize, that the geometric structure plays a key role for the evolution of potoexctiations in such systems [44]. This has been the topic of a further recent theoretical investigation [219]. Effects of chain conformation and mesoscopic morphology typically have a strong influence on the delayed formation of interchain polaron pairs [216]. Light absorption in a chromophore results in a photoexcitation with a Coulomb-bound electron-hole pair, which is delocalized over several repeating units [220] (compare section 2.4.3). Such a neutral photoexcitation is expected to ionize more efficiently into localized polaron pairs, if it can experience sites with imbalanced electronegativity, otherwise it localizes as an exciton. In donor–acceptor-conjugated co-polymers, sites with different electron affinity are distributed in an alternating fashion along the whole polymer chain and, potentially, each one can act as a localization site for an electron or a hole. Thus, a large difference in electron affinity between the donor and acceptor allows not only to obtain low bandgaps, as discussed in Figure 5.6 a) – d), but enhances also the formation yield of polaron pairs in copolymers, as shown in Figure 6.4 a) – d). This is demonstrated by PCPDT-BDT and PCPDT-2TTP, which both show high yields with a difference in η within our estimation errors. However, a noticeable decrease of the yield of about a factor of 1.3 is rather observed when, comparing copolymers with the same acceptor and with/without spacers. On one hand, this observation indicates that a smaller separation of donor and acceptor is beneficial for the formation of a polaron pair from the initial photoexcitation. Most likely this is due to a larger probability of the initial exciton to experience a localizing site.

On the other hand, donor-acceptor copolymers with closely spaced moieties suffer from fast recombination of the polaron pairs. This recombination is likely due to the close average separation of the correlated geminate pairs. We attribute the fast initial decay time to geminate on chain recombination of polaron pairs. The different separation between localized electron and hole polarons in polymers with and without spacers could be the reason for the larges differences of recombination times for copolymers with and without spacers. Our measurements point to a beneficial role of more extended structures in increasing the recombination lifetime (Figure 6.5). Here, the increased spatial separation may hinder the electron hole overlap, which is necessary for polaron pair recombination. In addition, an increased separation corresponds to a weaker Coulomb attraction, and a more probable escape from the electrostatic potential. This happens most likely by thermally activated hopping along the same chain or to adjacent chains. We note that the delocalization of one of the carriers, in particular holes, is crucial to avoid fast recombination and seems to be a promising approach for a structural optimization of polaron pair lifetime in donor-acceptor copolymers [44,219].

A further interesting finding is that stronger electron affinities of the acceptor lead to faster recombination times. Because of the energetically deeper $LUMO_{Acceptor}$, the electron polaron is expected to be more localized what in turn promotes the recombination. In fact, polaron pair formation is expected to break the symmetry of the π-orbital, with the electron localizing mainly on a orbitals of isolated acceptor LUMOs [221]. Thus, electrons in PCPDT-BDT and PCPDT-2TTP may have poor intrachain carrier mobilities, compared with co-polymers with $LUMO_{Acceptor}$ levels at higher energy. The long initial polaron pair decay time (> 12 ps) in P3HT supports our interpretation. Here, polaron pair formation is driven by exciton delocalization at sites with energetic disorder (typically > 50 meV) [222], since energetic disorder within the chromophore is absent in homopolymers. A recent work by Reid et al. [57] showed that the polaron yield in P3HT depends on the degree of microcrystallinity. Charge separation is believed to occur at the interface between crystalline and amorphous regions. A less pronounced carrier localization in one of the two regions can lead to large intrachain mobilities [218,223] and a longer polaron pair lifetime.

These results have immediate implications for organic photovoltaics. In photovoltaic devices, the long-living charge separation is accomplished by exciton dissociation at the interface between phase-separated domains, where electrons and holes can diffuse away. Conjugated copolymers with high yields in the formation of polaron pairs are expected to require fullerene acceptors with smaller LUMO energy offsets. A lower driving force might be sufficient for their dissociation, since polaron pairs have a lower binding energy than excitons, as discussed in section 2.4.3. The preference for copolymers with either long-living polaron pairs or large generation yields may depend on the morphology of the blends with the fullerene. Blend systems with small polymer domains might be beneficial for copolymers with high yields of formation. In this configuration, polaron pairs will be separated in free charges within their lifetime, as they will be likely generated directly at the interface or close to it. For large polymer domains (> 20 nm), a longer lifetime of the polaron pairs will be beneficial to favor diffusion to the interface.

These findings report important insights on how the chemical structure of copolymers is influencing the yield of generation and the recombination dynamics of polaron pairs. The generation yield depends on the electron affinity of the acceptor moiety, but also on on the separation between the centers of mass of donor and acceptor. The results might provide useful input into the understanding of structure-property relationships in low-bandgap polymers for photovoltaics.

6.4. Structural influences on exciton and polaron mobility in donor-acceptor copolymers

When talking about mobilities in polymers for organic electronics, like transistors or photovoltaic cells, the discussion is typically limited to the mobility of charge carriers being transported and extracted via the polymer, i.e. hole polarons. From the present literature, both exciton and charge carrier mobilities in polymer films are well known to depend crucially on the morphology[224]. For this reason, the values for hole polaron mobilities in conjugated polymers, found in literature, cover several orders of magnitude, i.e. 10^{-10} to 10^{-5} cm^{2}V^{-1}s^{-1} for PPV polymers [225,226], and $8.5 \cdot 10^{5}$ to 10^{3} cm^{2}V^{-1}s^{-1} for devices and blends with P3HT [227,228]. Even higher values, on the order of 0.1 cm^{2}V^{-1}s^{-1}, could be found in P3HT films, with a very high degree of cristallinity. The latter allows efficient inter-chain delocalization of hole polarons [31]. But not only morphological issues are known to play a role. A strong influence of only slight changes in their chemical structure was found in a systematic study of charge carrier mobilities in low-bandgap copolymers by

Scharber et al. [229]. By replacing the carbon bridging atom in the dithiophene unit of PCPDT-BT with a silicon atom, the hole polaron mobility increased by a factor two, from $5 \cdot 10^{-3}$ to 10^{-2} $cm^2V^{-1}s^{-1}$. Obviously, the molecular chemical structure has strong influence on the mobility of hole polarons, thus an effect on electron polaron mobility is straight forward to assume. Of course the mobility of separated charge carriers remains a crucial factor for their efficient extraction from a photovoltaic device. But nevertheless, charge dissociation at a donor-acceptor interface remains a fundamental step for their formation. Therefore, polaron pair mobility is a parameter of great interest, in order to understand and optimize their diffusion and subsequent extraction.

In literature, it is commonly implied that polaron pairs either undergo geminate recombination or dissociate at the place of their formation, as their diffusion is not considered in theoretical models [71,230,231]. Nevertheless, there are experimental indications that charge carriers formed in conjugated polymers exhibit a very high initial mobility [232] and are also capable of moving to other molecules or chromophores before undergoing recombination or dissociation [233]. Only recently, theoretical studies investigated the role of polaron pair diffusion [74] and came to the conclusion that it has to be taken into account for describing BHJ solar cells. They showed that diffusion of Coulomb-bound charge carriers is responsible for a severe loss channel. The latter becomes dominant at high electric fields and leads to a loss of more than 40% of the generated charges at the interface with an electrode [74].

This study reveals that polaron pair diffusion benefits from a high local mobility, which can exceed the average macroscopic mobility by far [74], and that polaron pairs can also have a very high initial mobility directly after their formation [232]. These findings have a strong implication for the applicability of polaron pair extraction in organic solar cells.

In order to proof and investigate the different character of excitons and polarons in donor-acceptor copolymers, two different materials have been investigated with various densities of photoexcitations. Varying the excitation density allows to change the average distance between photoexcitations. An estimate of the diffusion length and mobility of different species can be given, by measuring the onset of bimolecular recombination effects in dependence of their average separation distance.

For a density dependent comparison of exciton and polaron pair decay dynamics, the copolymers PCPDT-BT and PCPDT-2TBT have been chosen. In Figure 6.6 the decay dynamics of *GB*, *SE* and *P₁* are shown for PCPDT-BT (panels a) - c)) and PCPDT-2TBT (panels d) – f), when excited with different excitation densities at 760 nm and 640 nm, respectively.

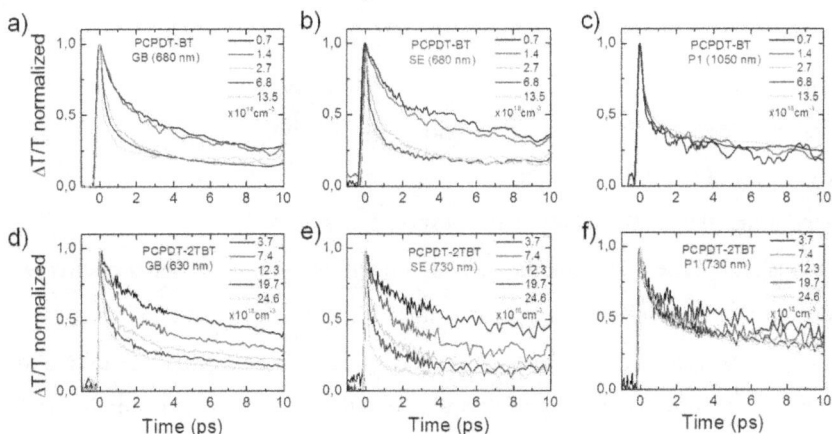

Figure 6.6: *Recombination dynamics on a 10 ps timescale for various photoexcitation densities of GB, SE and P1 of PCPDT-BT (panels a) - c)) and PCPDT-2TBT (panels d) – f)). For increasing excitation densities a significant shortening of the SE decay time is observed for both materials. The P₁ decay time of PCPDT-2TBT shows a less pronounced decrease, while no change can be seen for PCPDT-BT. The very similar dynamics of GB and SE in both materials point to the fact, that excitons are the major species formed upon photoexcitation with 760 nm and 640 nm, respectively.*

While *SE* exclusively originates in the photoluminescence of excitons, P_1 is an absorption band originating only in polarons. For better comparability, suitable excitation wavelengths have been chosen to address the maximum of the low-energy absorption band of each material. Within the first picoseconds, an initial rapid decay of polaron pairs occurs, which could not be observed to the same extent in previous measurements exciting at 800 nm and probing in the MIR. It is most likely originating in a higher excitation density, required for retrieving a full $\Delta T/T$ spectrum at moderate repetition rates, as explained in the experimental chapter, or due to a residual overlapping contribution of *SE*.

The respective densities are denoted as an inset in each panel. For PCPDT-BT a very strong dependence of the *SE* decay time is observed, as shown in panel b), decreasing rapidly with increasing photoexcitation density, till reaching saturation at a density of ca. $3 \cdot 10^{18}$ cm^{-3}. The latter corresponds to an average distance between two photoexcitations of 3.2 nm. In contrast, the decay dynamics of the polaron pairs, shown in panel c), seem not to be affected by an increasing excitation density in the observed range.

In this latter case, the slight differences are within the error of the measurements due to the relatively small signal to noise ratio for lower excitation densities. They show a rapid initial decay, followed by a decay on a longer timescale than the *SE*. *GB* shows a shorting of decay time with increasing density similar to *SE*. Saturation is again reached at a density of about $3 \cdot 10^{18}$ cm^{-3}. A comparable behavior can be found for PCPDT-2TBT, qualitatively. Also for this material, the *SE* band (panel e)) exhibits a clear shortening of its decay time with increasing excitation density, but saturation is observed only at higher densities of about $1.2 \cdot 10^{19}$ cm^{-3}. This corresponds to an average distance of only 2 nm between two photoexcitations. The polaron pair signal, plotted in panel f), is much less affected by an increasing excitation density than the *SE* band, but shows a slight decrease of lifetime at higher densities. Similar to the case of PCPDT-BT, the variation of the *GB* dynamics can be compared to the dynamics found for the *SE* signal. A successive shortening of the *GB* decay time is present for smaller distances between photoexcitations, reaching saturation at the same value as the *SE*.

A closer look is required to point out the influence of excitation density on polaron pair diffusion. For better comparability, the signal amplitudes of *GB*, *SE* and P_1 were renormalized by their values at zero delay, and are plotted for different excitation densities in Figure 6.7.

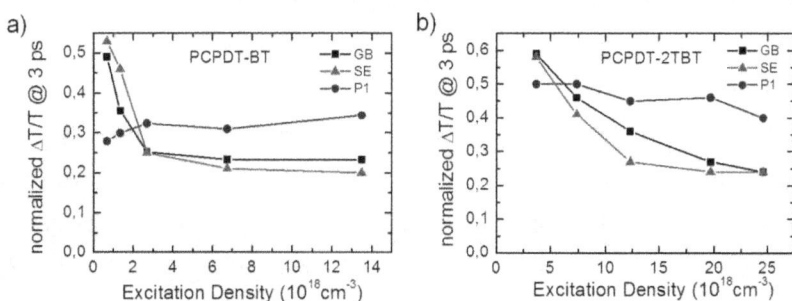

Figure 6.7: *Signal amplitudes of GB (black squares), SE (red triangles) and P_1 (blue dots) at 3 ps time delay renormalized by the respective zero-delay amplitude versus excitation density for a) PCPDT-BT and b) PCPDT-2TBT. PCPDT-BT shows a more sensitive decrease of the SE decay time than PCPDT-2TBT. In contrast, the P_1 decay time is clearly affected by excitation density only in -2TBT but not in –BT. The apparent increase in the latter case can be explained by a measurement artifact. The close similarity of SE and GB response indicates excitons being the major species of photoexcitations.*

In Figure 6.7 a) again the rapid decrease of exciton lifetime (red triangles) with increasing density can be observed, reaching saturation at a density of about $3 \cdot 10^{18}$ cm^{-3} for **PCPDT-BT**. The polaron signal (blue dots) does not show a detectable decrease with growing density, but seems to increase slightly. This can be considered as a measurement artifact due to a small spectral tail of overlapping *SE* at 1050 nm, where the polaron signal has been measured. With this visualization the similarity between *GB* (black squares) and *SE* is obvious. For PCPDT-2TBT, as shown in Figure 6.7 b), the weaker sensitivity of *SE* and *GB* decay time to the excitation density becomes clearly visible compared to PCPDT-BT. It exhibits a much slower drop-off and saturation is reached only at a significantly higher density of about $1.2 \cdot 10^{19}$ cm^{-3}, what obviously uncovers a different behavior compared to PCPDT-BT. Furthermore, Figure 6.7 b) reveals a decrease of lifetime for polarons with increasing density for. Going from a density of $3.7 \cdot 10^{18}$ to $2.5 \cdot 10^{19}$ cm^{-3}, the plotted signal ratio drops by about 20%, which clearly indicates a density dependent polaron lifetime for this material.

These findings demonstrate that in both materials excitons, which are the origin of *SE*, are much more affected by the excitation density (and hence the mean distance between two photoexcitations) than polaron pairs. This leads to the interpretation, firstly, that both are doubtlessly distinct species of photoexcitations and that both species can be characterized by different diffusion lengths and mobilities. Secondly, the slight affection of the polaron pair decay time in the -2TBT copolymer is in the contrary not present at all or below the detection threshold in –BT. The distinct exciton mobilities in these materials state different properties in terms of exciton- and polaron pair diffusion. Although these materials exhibit very similar chemical structures, saturation of bi-excitonic effects is reached at a mean distance between excitations of 3.2 nm in the case of –BT and only of 2 nm in –2TBT. On one hand, this finding indicates that the additional thiophene spacer in the -2TBT copolymer hinders exciton diffusion, most likely due to a reduction of mobility compared to the –BT copolymer. On the other hand, the presence of additional thiophene spacers in -2TBT favors the diffusion of polaron pairs, leading to a higher mobility and consequently a larger affection regarding their non-geminate (bi-molecular) recombination.

In conclusion, the differences in behavior of the exemplary copolymers PCPDT-BT and PCPDT-2TBT indicate that slight changes in the chemical structure have strong effects on their electronic properties. While the addition of thiophene spacers seems to hinder the mobility of excitons, a favorable effect on polaron pair mobility is found. This implies that different structural properties play a crucial role not only in terms of the formation yield, but also in terms of mobility of polaron pairs. The large range of variable parameters, regarding the chemical structure of materials, hopefully will allow in future to combine a high polaron pair formation yield with an optimized mobility.

6.5. Chapter summary

In this chapter, we report on enhanced polaron pair formation in donor-acceptor copolymers. Photoexcitation of thin polymer films without the addition of any electron accepting material, results mainly in the formation of strongly bound neutral excitons and to a certain extent of weakly bound polaron pairs. For homopolymers, like P3HT, this polaron pair yield η, also called polaron pair to exciton branching ratio, is found to be small ($\approx 8\%$) and to depend crucially on morphologically induced energetic disorder between amorphous regions and crystalline domains. In contrast, donor-acceptor copolymers exhibit much higher intrinsic yields of up to $\eta \approx 24\%$. We showed that the energetic disorder induced by alternating donor and acceptor moieties along the copolymer backbone favors the formation of polaron pairs in a very high degree. A suitable choice of donor-acceptor materials allowed a systematic investigation of structural correlations with the branching ratio and the lifetime of the formed polaron pairs. Strong electron affinity of the on-chain acceptor moieties is found to have a very beneficial influence on the formation yield of polaron pairs, but decreases their recombination time dramatically. A larger spatial extension of the donor unit and the subsequent larger separation of the donor and acceptor centers, by the incorporation of thiophene spacers, increase the recombination time, but lowers the yield of polaron pairs. The incorporation of thiophene spacers has also immediate consequences on the mobility of polaron pairs and excitons. A systematic study on similar copolymers, with and without spacers, at different photoexcitation densities indicates that their presence hinders exciton but favors polaron pair the mobility. These results further prove that polaron pairs and excitons in donor-acceptor copolymers are indeed distinct species of photoexcitations and can be distinguished by their recombination dynamics and their different affection to bi-molecular processes. Altogether, the results presented in this chapter demonstrate that the electronegativity and on-chain arrangement of donor and acceptor units has a strong influence on the formation and lifetime of polaron pairs. These correlations point to new degrees of freedom for a future optimization of such materials in terms of formation yield, lifetime and mobility of polaron pairs.

7. The beneficial role of excess energy and small molecule size for polaron pair formation

The results presented in the last chapter show that the chemical structure of donor-acceptor copolymers favors the formation of polaron pairs in the absence of fullerenes or other electron accepting compounds. Ultrafast spectroscopic measurements revealed a polaron pair to exciton branching ratio of more than 24% for these materials, exceeding homopolymers by a factor of 3 ($\approx$ 8%). A strong dependence of this yield on the on-chain topology and the acceptor electron-affinity was found. For those measurements, all polymers have been excited in the low energy band of the characteristic "camel back" absorption spectrum [44]. Theoretical studies by Jespersen et al. predict the formation of excited states with a substantial charge transfer character only for absorption in this low-energy band, while fully delocalized states are expected for excitation in high-energy bands [147]. Since this charge transfer character of excited states is believed to be essential for a high formation yield of polaron pairs, it remains a valid question whether a substantial yield can also be achieved for excitations with a significant amount of excess energy. An extraction of polaron pairs will only be applicable and of interest for organic photovoltaics, if a high polaron pair formation yield is present throughout the whole absorption spectrum of the favored material. This crucial question shall be addressed in this chapter by broadband ultrafast spectroscopic measurements with various excitation photon energies. In addition, the effect of molecule chain length shall be studied in order to further complete the investigations of influencing material parameters on the formation of weakly bound polaron pairs in donor-acceptor materials [166].

113

The beneficial role of excess energy and small molecule size for polaron pair formation

7.1. Excess excitation energy promoting polaron pair formation in donor-acceptor materials

Before going into experimental and theoretical details, this chapter shall focus on the dependence of localization and charge-transfer character of the populated orbitals on the evolution of photoexcitations in a compact and intuitive way. Figure 7.1 presents a simplified physical picture of polaron pair formation in donor-acceptor materials when absorbing photons with and without significant excess energy.

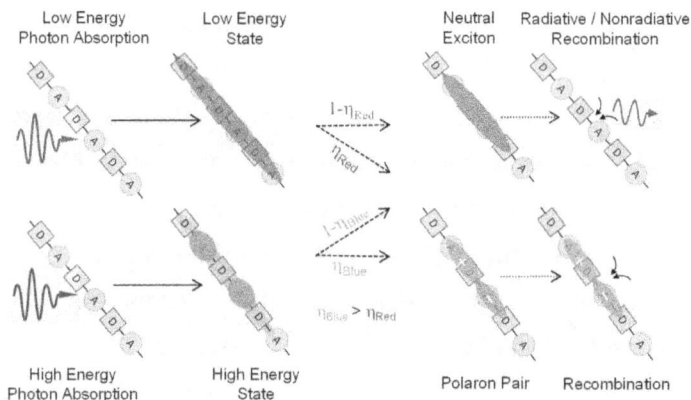

Figure 7.1: *Schematic illustration of the evolution of photoexcitations with different excess energies. Low energy excitations mainly result in the formation of delocalized states (upper part) which mainly undergo the evolution towards neutral excitons. In contrast, excitations with excess energy address orbitals with a high degree of charge-transfer and localization. These states favor the formation of spatially separated weakly bound polaron pairs. In all studied donor-acceptor materials, the found yield for high-energy excitation is higher than for excitations close to the bandgap, i.e.* $\eta_{Blue} > \eta_{Red}$.

In contrast to the expectations based on former theoretical studies [147], photoexcitations close to the optical bandgap result mainly in the occupation of orbitals, which are delocalized over the chromophore and do not exhibit noticeable charge-transfer character [166]. The resulting large wave function overlap and the strong Coulomb interaction of electron and hole mainly lead to the formation of neutral excitons and thus only to a small amount polaron pairs. These formed excitons provoke pronounced signatures of excitonic absorption and stimulated emission in the optically induced differential absorption spectra. Both are decaying on a timescale characteristic for excitons in these systems. In contrast, after photoexcitation with significant excess energy, the populated higher lying orbitals exhibit a much higher degree of charge-transfer and consequently further spatial separation of electron and hole. The reduced Coulomb attraction and pre-separation favors the formation of weakly bound polaron pairs and exhibits a higher polaron pair yield for photoexcitations with excess energy, i.e. $\eta_{Blue} > \eta_{Red}$. This can be observed by a reduced signal of the exciton species and a strengthened signal of polaron absorption with a decay time unique for polaron pairs in the respective system.

Bearing in mind this intuitive physical picture, we will investigate the influence of localization and charge-transfer by ultrafast spectroscopic studies and theoretical investigations of the involved orbitals in sections 7.2 and 7.3.

7.2. Ultrafast spectroscopic investigation of polaron pair formation with various excitation energies

In order to investigate the dependence of polaron pair formation on excess energy of the absorbed photon, a suitable set of materials was chosen. In addition to the already established copolymers PCPDT-BT and PCPDT-2TBT the oligomer 7,7'-(4,4-bis(2-ethylhexyl)-4H-cyclopenta[1,2-b:5,4-b']dithiophene-2,6-diyl)bis(4-methylbenzo[c][1,2,5]thiadiazole) (CPDT-BT) was selected for this investigation. The latter is a small molecule that shares the same chemical structure with the copolymer PCPDT-BT, but consists only of

116

one donor unit, capped by one acceptor unit on both sides. The respective chemical structures of all materials are plotted in Figure 7.2 a) – c), complete with the newly added small molecule.

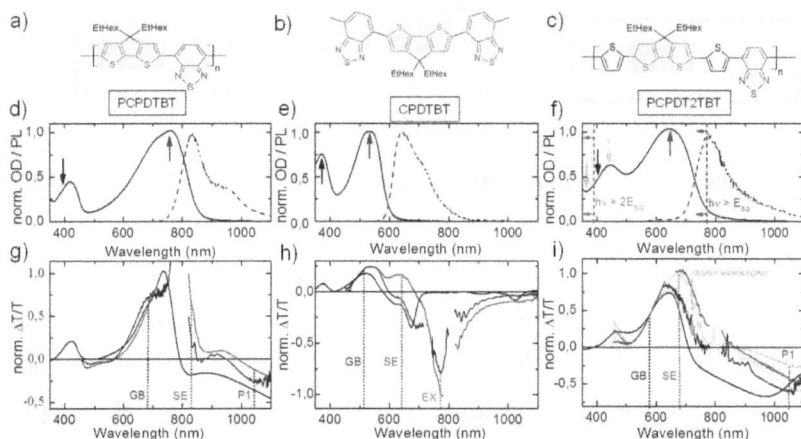

Figure 7.2: *a) – c): Chemcial structures of PCPDT-BT, CPDT-BT and PCPDT-2TBT. d) – f): show the absorptivity of neutral molecules in thin film samples with low- and high-energy excitation wavelength indicated as red and blue arrow, respectively. Further high energy excitations of PCPDT-2TBT are indicated as light blue and light green arrows. g) – i): Broadband ΔT/T, 350 fs after excitation for low- (red) and high-energy (blue) excitation, compared to the chemically induced -ΔOD of the respective cations (black). All curves are normalized to the ground state bleaching for better comparability. Vertical dotted lines indicate the spectral positions for evaluating the temporal dynamics of GB (black), SE (red), P_1 (blue) and EX (green).*

Only recently, a similar oligomer was used successfully in a small-molecule organic solar cell with a record power conversion efficiency of 6.7% [155]. In the framework of the present study, all materials were excited at the low- and high-energy absorption peaks of their related camel back absorption spectra, indicated by red and blue arrows, respectively. As shown together with the ground state absorption (solid black line) and photoluminescence spectra

117

(dashed black line) in Figure 7.2 d) – f), the suitable wavelengths for this purpose are 760 nm and 400 nm for PCPDT-BT, 530 nm and 370 nm for CPDT-BT, and 640 and 400 nm for PCPDT-2TBT. For the latter copolymer, an extended investigation with excitation wavelengths below and above twice the optical bandgap $h\nu_{Pump} > 2 \cdot E_G$ (vertical dashed green line) was performed. The purpose was to clarify the occurrence of singlet fission (SF), multi exciton generation (MEG), as it was partially observed for other organic materials, such as pentacene [234–237] or carbon nanotubes [238]. The other excitation wavelengths used for this purpose are 440 nm (light blue arrow) and 370 nm (light green arrow). For probing GB, SE, P_1 and EX simultaneously, a broadband white light was used, covering the spectral range from 450 nm to 780 nm and from 820 nm to 1100 nm. This allows measuring all spectral features at the same time and normalizing the obtained signals to the ground state bleaching for better relative comparability. The exciton absorption band EX, which is located as expected at longer wavelengths than P_1, is only detectable in the available spectral window for the oligomer because of its large optical bandgap due to the short chain- and conjugation length.

All shown ultrafast broadband $\Delta T/T$ spectra are taken 350 fs after excitation. When comparing both normalized $\Delta T/T$ signals for 760 nm and 400 nm excitation of PCPDT-BT in Figure 7.2 g), ground state bleaching is observable from 570 nm up to 745 nm in good agreement with the ground state absorptivity shown in panel a). Further in the infrared, peaking below 820 nm and exhibiting a shoulder at 935 nm, a second strong positive contribution is obvious. Because it matches well with the photoluminescence spectrum, an assignment to stimulated emission is straight forward, which partially overlaps on its blue edge with the GB. Comparing it with the chemically induced $-\Delta OD$ (black curve), a negative contribution of polaron absorption is obvious. The latter starts at about 800 nm and exceeds the detection window at 1100 nm with the blue edge being again covered by a positive SE signal. At 950 nm, the contribution of the polaron signal becomes dominant and the total signal changes its sign to negative. When excited with excess energy, a clearly less-pronounced *SE* and enhanced polaron absorption P_1 are strong indications for a higher branching ratio,   i.e.

$\eta_{Blue} > \eta_{Red}$.

For the oligomer CPDT-BT, shown in Figure 7.2 h), we observe a similar pattern of spectroscopic features with an additional absorption band peaking at 780 nm. Its larger optical bandgap and thus blue-shifted absorption spectrum compared to the related copolymer allows accessing its excitonic absorption band *EX*, located between 700 nm and 1100 nm. The reduced signal strength, when excited with excess energy clearly proves a reduced formation of excitons than for low-energy excitations. Due to the large overlap of *SE* and P_1 for the oligomer, the $\Delta T/T$ signal completely changes its sign around 630 nm from positive *SE* to negative P_1 signal, when exciting with significant excess energy. In qualitative agreement with the findings for the copolymer PCPDT-BT, the oligomer shows a larger branching ratio ($\eta_{Blue} > \eta_{Red}$) for excitation with excess energy and even exhibits a much larger effect than observed the polymer. This indication for a higher polaron pair yield is further confirmed by the above finding of a reduced excitonic absorption for short-wavelength excitation.

For PCPDT-2TBT, as shown in Figure 7.2 i), an increasing amount of excess energy (curves in the order of red, light blue, blue, light green) leads to a successive reduction of *SE* in the range from 640 nm to 860 nm, while the spectral shape of *GB* below 630 nm remains largely unaffected. Interestingly, the negative polaron absorption in the spectral range from 860 nm to 1100 nm is increasing monotonously for excitation wavelengths from 640 nm to 400 nm, but drops drastically by a factor of 3 when going from 400 nm to 370 nm. Up to 400 nm, the trend of decreasing *SE* and increasing P_1 signal agrees qualitatively with the phenomena found in both of the other investigated materials. Further, it shall be mentioned explicitly that *SE* remains absent also for excitation at 370

119

nm. This excitation wavelength corresponds to a photon energy twice exceeding the material's optical bandgap. The drastic drop of polaron signal together with the absence of *SE* can by no means be regarded as an indication for multi exciton- or multi-polaron pair formation. Most likely, the populated high-lying energetic state, opens a non-radiative (dark) loss channel, allowing relaxation without stimulated emission or polaron pair formation. The found absence of MEG is in good agreement with other recent studies on a model polymer system for photovoltaics by Bange et al. [239].

Further experimental proof of a higher branching ratio for photoexcitations with significant excess energy can be found by a detailed look at the temporal dynamics of the spectral features shown in Figure 7.3. Figure 7.3 shows the decay dynamics of all spectral features described in Figure 7.2. In the case of PCPDT-BT about 65% of the initial polaron pair signal (blue line) decays on an ultrafast timescale within the temporal resolution of the setup, followed by a slow decay with a time constant of 18 ps. We attribute this initial decay to a fast geminate recombination of on-chain polaron pairs since their decay dynamics are not significantly affected by a changing excitation density, as already commented in section 6.4. Previous measurements of polaron pair dynamics at even lower excitation density, probing the P_2 absorption band in the MIR, support this assumption [44]. However, using a broadband probe and a suitable detection system requires measuring with a repetition rate of only 1 kHz and therefore higher excitation densities are necessary to maintain a reasonable signal to noise ratio. Although the used excitation densities were chosen as low as possible, a small residual contribution of bimolecular exciton recombination cannot be excluded. Slight interference with overlapping *SE* might be a further contribution to the appearance of this ultrafast feature. For the decay of stimulated emission (red line), bi-exponential decay dynamics with 1.3 and 16.5 ps decay constants could be evaluated. Since the general temporary absence of molecules in the ground state provokes the signature of *GB* (black line), a contribution of exciton and polaron dynamics to its decay dynamics is straight forward. Again, the pronounced similarity between *GB* and *SE* decay dynamics for low-energy excitation proofs excitons to be the major species of photoexcitations for 760 nm excitation. For excitations with excess energy at 400 nm, the fast initial decay of polaron pairs is reduced to about 40% of the initial amplitude, as apparent in Figure 7.3 d). With excess energy, the decay of *GB* shows decay dynamics which are comparable to the decay of polaron pairs, suggesting that the latter are the major species formed by excitation with significant excess energy. In agreement with other studies [179], the stimulated emission is observed with a slower rise after excitation at a wavelength of 400 nm, which can be explained by vibrational relaxation of the excited state prior to photon emission.

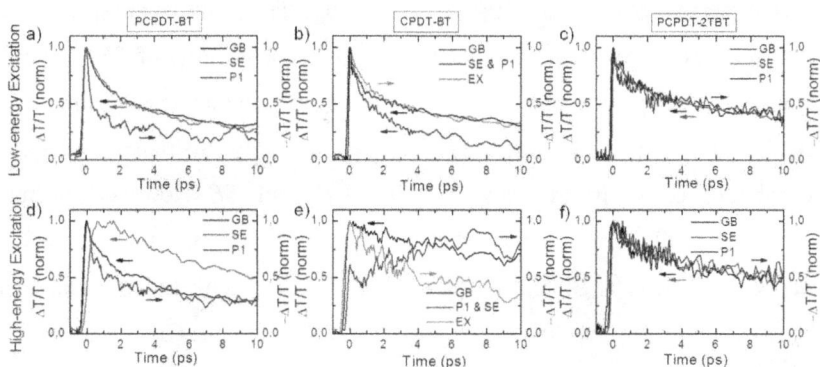

Figure 7.3: Ultrafast transient dynamics of GB (black), SE (red), P1 (blue) and EX (green) for long wavelength excitation of a) PCPDT-BT (760 nm) b) CPDT-BT (530 nm) and c) PCPDT-2TBT (640 nm). Their respective dynamics for short-wavelength excitation at 400 nm, 370 nm and 400 nm are shown below in panels d) – f). For better comparability all decay curves are normalized and eventually plotted with negative sign (as indicated in the graphs). Clear differences in the temporal dynamics between long- and short-wavelength excitation can be found for PCPDT-BT in going from panel a) to d). Those are

even more pronounced for the oligomer in panel b) and e). PCPDT-2TBT in panel c) and f) shows only small differences between the spectral signatures of GB, SE and P$_1$ and for various excitation wavelengths. Spectral positions of respective probe wavelengths are indicated by vertical dotted lines in Figure 7.2.

The temporal dynamics for the oligomer CPDT-BT are shown in Figure 7.3 b) and e) and confirm the hypothesis that excitation with excess energy leads to a higher branching ratio. While excitation close to the bandgap the *GB* decay follows almost exactly the dynamics of the exciton absorption band *EX* (green line), it appears to be clearly slower for excitation with excess energy, indicating a larger amount of polaron pairs among the photoexcitations. When excited at 530 nm (panel b)), the temporal dynamics in the spectral region from 600 nm to 700 nm, where *SE* and *P$_1$* are largely overlapping, clearly follow the shape of the photoluminescence spectrum, indicating a major contribution of *SE*. Although originating in the same species, *SE* shows faster recombination dynamics than *EX*, which most likely originates in the spectrally overlapping negative contribution of the polaron absorption band *P$_1$* with *SE*. However, the most interesting results are obtained when exciting with a large amount of excess energy, at a wavelength of 370 nm. In this case a long living signal with negative amplitude due to polaron absorption follows the *GB* dynamics and, according to our assignment, is another strong indication that polaron pairs are the mainly formed species. The seemingly slow buildup of the polaronic absorption band (panel e)) can be explained as the vanishing contribution of the overlapping *SE* signal due to a fast decay of excitons. Comparing the decay of polaron pairs in the oligomer and the related copolymer on a timescale of 200 ps, when excited with excess energy, a surprisingly large difference becomes obvious.

As shown in Figure 7.4, the polaron pairs in the oligomer exhibit a lifetime, which is more than one order of magnitude longer (164 ps) than the slow component of the bi-exponential decay in the copolymer (13 ps).

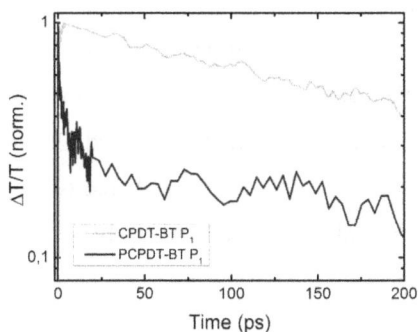

$\Delta T/T$ (norm.)

1

0,1

CPDT-BT P$_1$
PCPDT-BT P$_1$

0   50   100   150   200

Time (ps)

Figure 7.4: *Ultrafast dynamics of polaron pair absorption in CPDT-BT (light blue) and PCPDT-BT (dark blue) on a timescale of 200 ps. While the polaron pairs in the oligomer show a lifetimte of 164 ps when excited with excess energy, polaron pairs in the copolymer exhibit as a long component of their bi-exponential decay only a lifetime of 13 ps.*

For PCPDT-2TBT, as shown in Figure 7.3 c) and f), the differences in the temporal decay dynamics of polaron pairs, *SE* and *GB* are much smaller than for the other two materials. This holds true for excitations close to the bandgap (panel c)) and also for those with significant excess energy (panel f)). One possible answer for this close similarity of decay dynamics of both species might be their more leveled mobility compared to PCPDT-BT, as discussed in detail in section 6.4. Nevertheless, this finding does not contradict to a higher polaron pair formation yield for excitations with excess energy, as is also observed for this polymer in the *ΔT/T* spectra shown in Figure 7.2 i).

In summary, all materials show a higher polaron-pair yield η for excitation with higher photon energies compared to excitations close to the optical bandgap, i.e. $\eta_{Blue} > \eta_{Red}$. This effect is well observable for the donor-acceptor copolymers PCPDT-BT and PCPDT-2TBT and is even significantly more pronounced for the shorter oligomer CPDT-BT. The latter finding is a strong indication that the molecular chain length has significant influence on the evolution of photoexcited states in donor-acceptor materials, as well as on the lifetime of the generated polaron pairs.

In order to reveal the nature of this influence and the role of charge localization and charge-transfer character, quantum chemical simulations were performed [166]. They allow an investigation of the involved electronic states addressed with different excitation energies. A detailed discussion of these theoretical investigations and their findings are presented in the following section.

7.3. The key role of charge-transfer character and localization of photoexcited states for polaron pair formation

To obtain a deeper understanding and to provide a more comprehensive explanation on the nature of polaron pair formation and the influence of excess excitation energy, an investigation by quantum chemical modeling of the involved optical transitions was performed [166]. Details about the computational methods can be found in section 4.3.4. In order to shed light on the role of molecular chain length and conjugation length in the process of polaron pair formation, the copolymer PCPDT-BT and the related monomer CPDT-BT were chosen for a close comparison. Figure 7.5 provides natural transition orbitals (NTOs) for the relevant optical transitions in PCPDT-BT (panel a)) and CPDT-BT (panel b)).

Each NTO pair has two parts: The lower part, with the initial electron orbital, and the upper part, with the final orbital, which is reached after optical excitation. In a figurative sense, the first can be regarded as the left-over hole wave function while the latter one corresponds to the final electron wave function. The given wavelength value corresponds to the resonant excitation wavelength, whereas f denotes the transition's oscillator. As already discussed in section 5.2.2, the observed blue-shift of calculated wavelength compared to measured excitation wavelengths originates in the limited length of the calculated molecules and the used computational method.

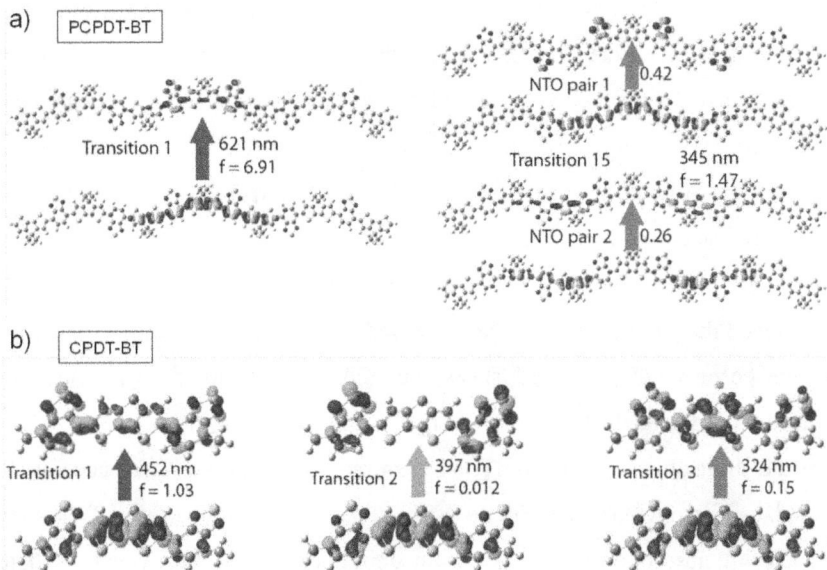

Figure 7.5: Natural transition orbitals (NTOs) governing the main absorption bands of the camel back absorption spectrum of a) PCPDT-BT and b) CPDT-BT. Values for the resonant excitation wavelength of each transition are given together with the respective oscillator strength. While low-energy excitation (red arrows) mainly results in transitions between delocalized states, excess excitation energy (blue arrows) leads mainly to transitions with a pronounced charge-transfer character.

In the case of the copolymer, excitation close to the optical bandgap results in a transition (panel a), transition 1) where both electron and hole wave function are delocalized over several BT and dithiophene units in the center of the chromophore. This strong wave function overlap justifies the large oscillator strength of transition *1* [47]. Furthermore, this large overlap and the common center of mass lead to a strong Coulomb interaction and explain the favored formation of neutral excitons instead of spatially separated polaron pairs.

After excitation with significant excess energy, i.e. in the high-energy absorption maximum, different NTO pairs are involved. Transition *15* carries the main oscillator strength for an excitation at 345 nm and consists of two NTO pairs (with CI coefficients of 0.42 and 0.26), as shown in Figure 7.5 a). A clear charge-transfer character is observed for both NTO pairs. At the dominant contributions, the electron wave function is mostly localized on the BT units, whereas the hole wave function is delocalized in nature. This spatial separation of electron and hole results in a reduced Coulomb interaction and consequently favors the formation of a weakly bound polaron pair. At this point it shall be mentioned that almost identical NTOs were found in the equilibrium geometries of the corresponding excited states, i.e. the final chain configuration after relaxation of the nuclei to the new energetic equilibrium position.

For the short oligomer CPDT-BT, an even more pronounced difference can be found, as it is obvious from Figure 7.5 b). Transition *1* is the origin of the long wavelength absorption band, arises from an initial and final state, which are both delocalized over the whole molecular backbone. Electron and hole are both calculated to be spread out over the whole molecule with their centers of mass located in the center of the molecule. The high-energy absorption band of CPDT-BT arises in transition *3*, which is only accessible with significant excess excitation energy. This transition appears to be delocalized with large overlap of the contributing electron and hole wave functions. However, following excitation of these transitions, we assume that the system undergoes ultrafast relaxation through several excited states, in agreement with previous studies [179]. In the course of this relaxation, a transit through another excited state with a very pronounced charge-transfer character is very likely. One hot candidate is the upper orbital of transition *2* in Figure 7.5 b), which is a dark state and hence optically not directly accessible, but exhibits a high degree of charge-transfer character when populated during the relaxation from a higher lying excited state. It is no surprise that the excited state exhibiting a high degree of charge-transfer is a dark state, since its oscillator strength suffers severely from lacking wave function overlap [47].

The exceptionally large change in the branching ratio, as well as the long polaron pair lifetime in the oligomer, exceeds the respective value in the copolymers by far. This leads to the assumption that the limited dimensions of the oligomer chains also add an additional degree of localization to the wave functions in the oligomer what favors polaron pair formation upon excitation with excess energy. One possible explanation could also be the contribution of inter-molecular effects in oligomer films due to eventual close spacing between donor and acceptor units of adjacent molecules in all three dimensions. In the case of favorable packing and orientation, the formation of stabilized long-lived polaron pairs might be favored, as it was observed also at interfaces in copolymer heterojunctions [240].

The finding of enhanced charge-transfer character and localization of the electron wave functions after photoexcitations with excess energy clearly contradicts the expectations of other theoretical investigations published so far [147]. Nevertheless, the understanding of the nature of photoexcited states at higher energies achieved by the quantum chemical modeling presented in this work is capable of giving an explanation for the experimentally found higher polaron pair yield in this regime. Altogether, these findings are clear indications that a high polaron pair formation yield is present over the whole absorption spectrum of donor-acceptor materials and is even increasing at shorter excitation wavelengths. The theoretically gained insight that charge localization, chain length and possibly inter-molecular effects play a key role in polaron pair formation gives further directions towards the future optimization of materials for more efficient solar cells based on the extraction of weakly bound polaron pairs.

7.4. Chapter summary

In this chapter we report on the beneficial role of excess excitation energy and short molecular chain lengths for polaron pair formation in donor-acceptor materials. Ultrafast spectroscopic measurements with different excitation wavelengths, in combination with broadband spectrally resolved probe pulses, retrieve excitons the major formed species formed by photoexcitations in the low-energy absorption band. In contrast, photoexcitations with a significant amount of excess energy are found to favor polaron pair formation with higher yields than excitations close to the optical bandgap, i.e. $\eta_{Blue} > \eta_{Red}$. This effect present in both studied copolymers PCPDT-BT and PCPDT-2TBT. Qualitatively the same, but on a much higher level, this effect is observed also in the short oligomer CPDT-BT, sharing the same chemical structure with PCPDT-BT. For this material, an increasing amount of excess energy almost completely changes the evolution of photoexcitation from excitons to polaron pair formation. Also the lifetime for short wavelength excitation of the short oligomer is reported to be more than order of magnitude longer than for the long copolymer.

To complete the experimental investigation, excitations with energies twice exceeding the optical bandgap have been performed for PCPDT-2TBT. In agreement with other studies, no indications for singlet fission or multi exciton generation could be found. With such high excess energies, a very low amount of SE in combination with a drastically dropped polaron signal is presented, which points toward an open non-radiative decay channel during relaxation.

Enhanced theoretical studies of the involved natural transition orbitals clearly indicate the favorable role of higher lying energetic orbitals, which are addressed by photoexcitations bearing excess energy. Their pronounced charge-transfer character and the resulting reduction of spatial overlap between electron and hole lead to reduced Coulomb interaction and are thus believed to promote polaron pair formation in a very efficient way. Further localization due to restricted chain length is supposed to play an important role towards charge separation, as it is likely to explain the more pronounced effect in the oligomer. Furthermore, small molecule size and eventual favorable arrangement with neighbor molecules might lead to an eventual contribution of inter-molecular effects.

In conclusion, our findings presented in this chapter clearly demonstrate that a high degree of polaron pair formation is present over the whole absorption spectrum of these materials and even becomes more efficient with increasing excess energy. Furthermore, the results point to an advantageous role of limited chain lengths in small molecules compared to copolymers in terms of polaron pair formation. This fact implies an intrinsic advantage of small molecules toward photovoltaic applications, which might be worth a closer investigation in future.

8. Conclusions and outlook

The aim of this work was the investigation of the photophysical properties of a novel class of materials, i.e. conjugated donor-acceptor copolymers. Despite their widespread use in organic solar cells with record braking efficiencies, their properties in terms of light absorption and charge carrier generation on a molecular scale remained fairly unexplored so far. Due to their unique chemical structure with alternated donor and acceptor segments along the copolymer backbone a different behavior compared to homopolymers was expected.

In this study, we applied a large variety of experimental and theoretical methods to investigate the evolution of photoexcitations and charge carrier formation in donor-acceptor materials. For this purpose, a suitable set of materials was chosen, consisting of the donor-acceptor copolymer PCPDT-BDT, PCPDT-BT, PCPDT-2TTP and PCPDT-2TBT and the small molecule CPDT-BT. With steady state optical spectroscopic measurements and chemical ionization, the signatures of polarons in the infrared spectral region could be determined for each individual investigated material. Thorough theoretical studies on the absorption spectra of positive and negative polarons were carried out for the homopolymer PCPDT and the above mentioned copolymers. The observed absorption symmetry of positive and negative polarons in homopolymers turned out to be broken in donor-acceptor materials. Lacking symmetry between HOMO and LUMO orbitals was found as the reason for strongly distinct absorption spectra of electron- and hole-polarons. A further comparison of different copolymers revealed a strong influence of the chemical structure on the wave function overlap of the occupied orbitals and the contribution of different inter-and intra-band transitions to the lowest optical transitions of polaron absorption. Combining these experimental and theoretical results on cation and anion absorption, allowed us to determine the polaron pair absorption cross

130

section for each individual material. With this knowledge, a further quantitative analysis of spectroscopic measurements of all investigated copolymers was possible.

By using ultrafast pump-probe spectroscopy, exciting the copolymers in their low-energy absorption band, and probing with suitable infrared wavelengths, their photophysical response could be studied in realtime. These measurements revealed that upon light absorption not only strongly bound excitons are formed, but also a significant amount of spatially separated and thus weaker bound polaron pairs. The quantum yield for polaron pair formation in pristine films of these materials turned out to be up to 24%, which is remarkably up to three times higher than it is found in the homopolymer P3HT. Comparison of the different copolymers showed that their chemical structure further has a strong influence on the polaron pair formation. Acceptor units with a strong electron affinity exhibit a high polaron pair formation yield but reduce the recombination time significantly. Furthermore, a larger separation between the centers of donor and accpetor unit was found to reduce the polaron pair yield from 24 to 15% for copolymers with the same acceptor unit, but to have a benefitial effect in terms of an extended polaon pair lifetime. Intensity dependent studies on the recombination dynamics of polaron pairs and excitons revealed that both are well distinct species of photoexcitations and exhibit very different mobilities in thin films. While the mobility of neutral excitons exceeds the one of polaron pairs, both are strongly affected by the chemical molecular structure. The discovered intensity dependencies imply that an insertion of thiophene spacers between donor and acceptor units reduces exciton and promotes polaron pair mobility. Thus, a specially designed chemical structure might allow an optimization of copolymers towards a high formation yield, feasible lifetime and a considerable mobility of polaron pairs. With these degrees of freedom, the exctaction of polaron pairs instead of strongly bound excitons might become applicable in future. This would promise a reduced loss of voltage which is required for the dissociation of photogenerated charge carriers.

An important question regarding the applicability of polaron pair extraction remained, whether their yield is substantial for light absorption through the whole absorption spectrum. In order to shed light on this aspect, we studied the role of excess energy on the polaron pair formation yield. It was found that photoexcitations with a significant amount of excess energy exhibit a higher polaron pair formation yield than those close to the optical. The same effect was shown to be present in different copolymers and to be even more pronounced in a short oligomer of very similar chemical structure, i.e. CPDT-BT. The origin of the high polaron pair formation for excitations with excess energy was explained with a strong charge-transfer character of the involved high energetic orbitals. In contrast to low-energy excitation, their electron wavefunction is almost exclusively localized at several acceptor units, reducing the spatial overlap and thus the Coulomb interaction with the homogeneously delocalized hole. The significantly larger effect found for the short oligomer is of great interest in terms of advantages and disadvantages of small molecules compared to copolymers. In this material an increasing amount of excess energy completely turns the evolution of photoexcitations toward polaron pair formation. This might be due its limited chain length, which further restricts the spatial extension of the electron wave function. A contribution of inter-molecular effects between neighboring molecules due to eventual favorable stacking might be a further explanation for this pronounced effect. Clarification of this question will be the topic of future investigations.

In conclusion, the results presented in this work illustrate the important role of ultrafast spectroscopy for understanding the governing factors of excited state evolution and charge carrier formation in organic materials. This allowed to provide a detailed insight into the photophysical properties of donor-acceptor copolymers and oligomers for photovoltaics. Due to their unique chemical structure, with alternating donor and acceptor moieties, they do not only offer an extended absorption in the near infrared spectral region, but also a remarkably high yield of weakly bound charge carrier pairs. Since this yield is now known to be substantial through the whole absorption spectrum and proper understanding of the structural correlations could be obtained, the applicability of their extraction seems to be feasible. Specific optimization of materials towards a high yield of polaron pairs might allow a reduction of the driving force and hence the voltage loss, which is required for the separation of bound charge carrier pairs. A beneficial combination of copolymers with different acceptor materials for photovoltaic blends with a less pronounced LUMO level offset might be reached in future. Further future experiments revealing the dissociation efficiency in dependence of the LUMO offset between the copolymer and the acceptor might be able to give an estimation of the achievable reduction of voltage loss and by that guide the way towards promising material combinations for highly efficient organic solar cells.

Abbreviations

Abbreviation	Full name
BBO	β barium borate
BHJ	bulk heterojunction
CCD	charge-coupled device
CIGS	copper indium gallium (di)selenide
CMS	charge modulation spectroscopy
CPDT-BT	7,7'-(4,4-bis(2-ethylhexyl)-4H-cyclopenta[1,2-b:5,4-b']dithiophene-2,6-diyl)bis(4-methylbenzo[c][1,2,5]thiadiazole)
CT	charge transfer
DFT	density functional theory
DIBAH	diisobutylaluminium hydride
DMC	dimethylcobaltocene
EX	exciton absorption
FF	fill factor
FTS	fluoroalkyl trichlorosilane
FWHM	full width at half maximum
GB	ground state bleaching
GDD	group delay dispersion
HOMO	highest occupied molecular orbital
IHJ	interdigitated heterojunction
IRF	instrument response function
ISC	inter system crossing
ITO	indium tin oxide
LUMO	lowest unoccupied molecular orbital
MCT	mercury cadmium tellurite

MDMO-PPV	poly(2-methoxyl-5-(3,7-dimethyloctyloxy)-para-phenylene-vinylene)
MEG	multi exciton generation
MIR	mid infrared
NIR	near infrared
NTO	natural transition orbital
ODCB	ortho-dichlorobenzene
OLED	organic light emitting diode
OPA	optical parametric amplifier
P3HT	poly(3-hexylthiophene)
PCBM	[6,6]-phenyl-C_{61}-butyric acid methyl ester
PCPDT-2TBT	poly(4,4'-bis-(2-ethylhexyl)-4H-cyclopenta[2,1-b;3,4-b']-dithiophene-4,7-bis(2-thienyl)-2,1,3-benzothiadiazole)
PCPDT-2TTP	poly(4,4'-bis-(2-ethylhexyl)-4H-cyclopenta[2,1-b;3,4-b']-dithiophene- 2,3-diphenyl-5,7-bis(2-thienyl)thieno[3,4-b]pyrazine)
PCPDT-BDT	poly(4;4'-bis-(2-ethylhexyl)-4H-cyclopenta[2,1-b;3,4-b']-dithiophene benzo-[1,2-c;4,5-c']bis[1,2,5]thiadiazole])
PCPDT-BT	poly(2,6-(4,4-bis(2-ethylhexyl)-4H-cyclopenta[1,2-b;3,4-b']dithiophene)-4,7-benzo[2,1,3]thiadiazole))
PEDOT:PSS	poly-ethylene-dioxythiophene:polystyrenesulfonic acid
PHJ	planar heterojunction
PIA	photoinduced absorption
PL	photoluminescence
PPPV	poly(phenylenephenylenevinylene)
SE	stimulated emission
SF	singlet fission
TCO	transparent conducting oxide
TDDFT	time dependent density functional theory
THF	tetrahydrofuran
UV	ultraviolet
VIS	visible

Bibliography

[1]   F. Alt, *Die Sonne schickt uns keine Rechnung: Neue Energie, neue Arbeit, neue Mobilität*, 3rd ed., Piper Taschenbuch (2009).

[2]   "Renewables 2012 Global Status Report," REN21 Secretariat (2012).

[3]   "Energie-Mix mit viel Sonnenstrom: Anteil von Solarstrom auf Rekordwert gestiegen," *Focus Online* (2012).

[4]   C. J. Brabec and J. R. Durrant, "Solution-Processed Organic Solar Cells," *MRS Bulletin* **33**(07), 670–675 (2008) [doi:10.1557/mrs2008.138].

[5]   "NREL: National Center for Photovoltaics," <http://www.nrel.gov/ncpv/> (11 December 2012).

[6]   T. M. Clarke and J. R. Durrant, "Charge Photogeneration in Organic Solar Cells," *Chemical Reviews* **110**(11), 6736–6767 (2010) [doi:10.1021/cr900271s].

[7]   H. Yersin, *Highly Efficient OLEDs with Phosphorescent Materials*, 1. Auflage, Wiley-VCH Verlag GmbH & Co. KGaA (2007).

[8]   B. Kippelen and J.-L. Brédas, "Organic photovoltaics," *Energy Environ. Sci.* **2**(3), 251–261 (2009) [doi:10.1039/B812502N].

[9]   V. Dyakonov, J. Parisi, and N. S. Sariciftci, *Organic Photovoltaics: Concepts and Realization*, Softcover reprint of hardcover 1st ed. 2003, Springer Berlin Heidelberg (2010).

[10]   D. Veldman, O. İpek, S. C. J. Meskers, J. Sweelssen, M. M. Koetse, S. C. Veenstra, J. M. Kroon, S. S. van Bavel, J. Loos, et al., "Compositional and Electric Field Dependence of the Dissociation of Charge Transfer Excitons in Alternating Polyfluorene Copolymer/Fullerene Blends," *J. Am. Chem. Soc.* **130**(24), 7721–7735 (2008) [doi:10.1021/ja8012598].

[11]   M. Hallermann, I. Kriegel, E. Da Como, J. M. Berger, E. von Hauff, and J. Feldmann, "Charge Transfer Excitons in Polymer/Fullerene Blends: The Role of Morphology and Polymer Chain Conformation," *Advanced Functional Materials* **19**(22), 3662–3668 (2009) [doi:10.1002/adfm.200901398].

[12]   A. A. Bakulin, A. Rao, V. G. Pavelyev, P. H. M. van Loosdrecht, M. S. Pshenichnikov, D. Niedzialek, J. Cornil, D. Beljonne, and R. H. Friend, "The Role of Driving Energy and Delocalized States for Charge Separation in Organic Semiconductors," *Science* **335**, 1340–1344 (2012) [doi:10.1126/science.1217745].

[13] D. Herrmann, S. Niesar, C. Scharsich, A. Köhler, M. Stutzmann, and E. Riedle, "Role of Structural Order and Excess Energy on Ultrafast Free Charge Generation in Hybrid Polythiophene/Si Photovoltaics Probed in Real Time by Near-Infrared Broadband Transient Absorption," *J. Am. Chem. Soc.* **133**(45), 18220–18233 (2011) [doi:10.1021/ja207887q].

[14] F. Deschler, E. Da Como, T. Limmer, R. Tautz, T. Godde, M. Bayer, E. von Hauff, S. Yilmaz, S. Allard, et al., "Reduced Charge Transfer Exciton Recombination in Organic Semiconductor Heterojunctions by Molecular Doping," *Phys. Rev. Lett.* **107**(12), 127402 (2011) [doi:10.1103/PhysRevLett.107.127402].

[15] M. Yan, L. J. Rothberg, F. Papadimitrakopoulos, M. E. Galvin, and T. M. Miller, "Spatially indirect excitons as primary photoexcitations in conjugated polymers," *Phys. Rev. Lett.* **72**(7), 1104–1107 (1994) [doi:10.1103/PhysRevLett.72.1104].

[16] E. L. Frankevich, A. A. Lymarev, I. Sokolik, F. E. Karasz, S. Blumstengel, R. H. Baughman, and H. H. Hörhold, "Polaron-pair generation in poly(phenylene vinylenes)," *Phys. Rev. B* **46**(15), 9320–9324 (1992) [doi:10.1103/PhysRevB.46.9320].

[17] B. Schweitzer, V. I. Arkhipov, and H. Bässler, "Field-induced delayed photoluminescence in a conjugated polymer," *Chemical Physics Letters* **304**(5–6), 365–370 (1999) [doi:10.1016/S0009-2614(99)00260-2].

[18] M. Svensson, F. Zhang, O. Inganäs, and M. R. Andersson, "Synthesis and properties of alternating polyfluorene copolymers with redshifted absorption for use in solar cells," *Synthetic Metals* **135–136**(0), 137–138 (2003) [doi:10.1016/S0379-6779(02)00552-0].

[19] N. Blouin, A. Michaud, and M. Leclerc, "A Low-Bandgap Poly(2,7-Carbazole) Derivative for Use in High-Performance Solar Cells," *Advanced Materials* **19**(17), 2295–2300 (2007) [doi:10.1002/adma.200602496].

[20] T. L. Brown, H. E. LeMay, and B. E. Bursten, *Chemie: Studieren kompakt*, Pearson Deutschland GmbH (2011).

[21] C. E. Mortimer and U. Müller, *Chemie: Das Basiswissen der Chemie*, Thieme (2003).

[22] "http://www.chem.ucla.edu/harding/IGOC/H/hybrid_orbital.html."

[23] H. Haken and H. C. Wolf, *Molekülphysik und Quantenchemie: Einführung in die experimentellen und theoretischen Grundlagen*, 4., völlig neu bearb. u. erw. Aufl., Springer Berlin Heidelberg (2002).

[24] S. Haneder, "Correlation Between Electronic Structure and Light Emission Properties in Phosphorescent Emitters," Ludwig-Maximilians Universität München (2009).

[25] "Wikipedia."

[26] W. P. Su, J. R. Schrieffer, and A. J. Heeger, "Solitons in Polyacetylene," *Phys. Rev. Lett.* **42**(25), 1698–1701 (1979) [doi:10.1103/PhysRevLett.42.1698].

[27] A. J. Heeger, S. Kivelson, J. R. Schrieffer, and W.-P. Su, "Solitons in conducting polymers," *Rev. Mod. Phys.* **60**(3), 781–850 (1988) [doi:10.1103/RevModPhys.60.781].

[28] H. Sirringhaus, "Device Physics of Solution-Processed Organic Field-Effect Transistors," *Advanced Materials* **17**(20), 2411–2425 (2005) [doi:10.1002/adma.200501152].

[29] W. Brütting, S. Berleb, and A. G. Mückl, "Device physics of organic light-emitting diodes based on molecular materials," *Organic Electronics* **2**(1), 1–36 (2001) [doi:10.1016/S1566-1199(01)00009-X].

[30] C. J. Brabec, S. Gowrisanker, J. J. M. Halls, D. Laird, S. Jia, and S. P. Williams, "Polymer–Fullerene Bulk-Heterojunction Solar Cells," *Advanced Materials* **22**(34), 3839–3856 (2010) [doi:10.1002/adma.200903697].

[31] H. Sirringhaus, P. J. Brown, R. H. Friend, M. M. Nielsen, K. Bechgaard, B. M. W. Langeveld-Voss, A. J. H. Spiering, R. a. J. Janssen, E. W. Meijer, et al., "Two-dimensional charge transport in self-organized, high-mobility conjugated polymers," *Nature* **401**(6754), 685–688 (1999) [doi:10.1038/44359].

[32] W. Barford, *Electronic and Optical Properties of Conjugated Polymers*, Oxford University Press, U.S.A. (2009).

[33] W. Kohn and L. J. Sham, "Self-Consistent Equations Including Exchange and Correlation Effects," *Phys. Rev.* **140**(4A), A1133–A1138 (1965) [doi:10.1103/PhysRev.140.A1133].

[34] W. Kohn, "Nobel Lecture: Electronic structure of matter—wave functions and density functionals," *Rev. Mod. Phys.* **71**(5), 1253–1266 (1999) [doi:10.1103/RevModPhys.71.1253].

[35] M. A. L. Marques and E. K. U. Gross, "Time-Dependent Density Functional Theory," *Annual Review of Physical Chemistry* **55**(1), 427–455 (2004) [doi:10.1146/annurev.physchem.55.091602.094449].

[36] D. M. Ceperley and B. J. Alder, "Ground State of the Electron Gas by a Stochastic Method," *Phys. Rev. Lett.* **45**(7), 566–569 (1980) [doi:10.1103/PhysRevLett.45.566].

[37] C. Fiolhais, F. Nogueira, and M. A. L. Marques, Eds., *A Primer in Density Functional Theory*.

[38] E. K. U. Gross and W. Kohn, "Time-Dependent Density-Functional Theory," in *Advances in Quantum Chemistry* **Volume 21**, Per-Olov Löwdin, Ed., pp. 255–291, Academic Press (1990).

[39] G. Onida, L. Reining, and A. Rubio, "Electronic excitations: density-functional versus many-body Green's-function approaches," *Rev. Mod. Phys.* **74**(2), 601–659 (2002) [doi:10.1103/RevModPhys.74.601].

[40] E. Runge and E. K. U. Gross, "Density-Functional Theory for Time-Dependent Systems," *Phys. Rev. Lett.* **52**(12), 997–1000 (1984) [doi:10.1103/PhysRevLett.52.997].

[41] J.-L. Brédas, D. Beljonne, V. Coropceanu, and J. Cornil, "Charge-Transfer and Energy-Transfer Processes in π-Conjugated Oligomers and Polymers: A Molecular Picture," *Chem. Rev.* **104**(11), 4971–5004 (2004) [doi:10.1021/cr040084k].

[42] C. Risko, M. D. McGehee, and J.-L. Brédas, "A quantum-chemical perspective into low optical-gap polymers for highly-efficient organic solar cells," *Chemical Science* **2**(7), 1200 (2011) [doi:10.1039/c0sc00642d].

[43] S. Schumacher, A. Ruseckas, N. A. Montgomery, P. J. Skabara, A. L. Kanibolotsky, M. J. Paterson, I. Galbraith, G. A. Turnbull, and I. D. W. Samuel, "Effect of exciton self-trapping and molecular conformation on photophysical properties of oligofluorenes," *The Journal of Chemical Physics* **131**(15), 154906–154906–8 (2009) [doi:10.1063/1.3244984].

[44] R. Tautz, E. Da Como, T. Limmer, J. Feldmann, H.-J. Egelhaaf, E. von Hauff, V. Lemaur, D. Beljonne, S. Yilmaz, et al., "Structural correlations in the generation of polaron pairs in low-bandgap polymers for photovoltaics," *Nature Communications* **3**, 970 (2012) [doi:10.1038/ncomms1967].

[45] C. Wiebeler, R. Tautz, J. Feldmann, E. von Hauff, E. Da Como, and S. Schumacher, "Spectral signatures of polarons in conjugated co-polymers," *Journal of Physical Chemistry B, in press* [doi:10.1021/jp3084869].

[46] J. Müller, "Exzitonentransfer und Dissoziationsdynamik in konjugierten Polymeren," Ludwig-Maximilians Universität München (2003).

[47]   G. Lanzani, *The Photophysics behind Photovoltaics and Photonics*, John Wiley & Sons (2012).

[48]   M. Schwoerer and H. C. Wolf, *Organische Molekulare Festkörper*, John Wiley & Sons (2012).

[49]   P. Schrögel, A. Tomkevičienė, P. Strohriegl, S. T. Hoffmann, A. Köhler, and C. Lennartz, "A series of CBP-derivatives as host materials for blue phosphorescent organic light-emitting diodes," *J. Mater. Chem.* **21**(7), 2266–2273 (2011) [doi:10.1039/C0JM03321A].

[50]   S. Haneder, E. Da Como, J. Feldmann, J. M. Lupton, C. Lennartz, P. Erk, E. Fuchs, O. Molt, I. Münster, et al., "Controlling the Radiative Rate of Deep-Blue Electrophosphorescent Organometallic Complexes by Singlet-Triplet Gap Engineering," *Advanced Materials* **20**(17), 3325–3330 (2008) [doi:10.1002/adma.200800630].

[51]   R. Hull, *Properties of Crystalline Silicon*, IET (1999).

[52]   S. F. Alvarado, P. F. Seidler, D. G. Lidzey, and D. D. C. Bradley, "Direct Determination of the Exciton Binding Energy of Conjugated Polymers Using a Scanning Tunneling Microscope," *Phys. Rev. Lett.* **81**(5), 1082–1085 (1998) [doi:10.1103/PhysRevLett.81.1082].

[53]   H. Bässler, V. I. Arkhipov, E. V. Emelianova, A. Gerhard, A. Hayer, C. Im, and J. Rissler, "Excitons in π-conjugated polymers," *Synthetic Metals* **135–136**(0), 377–382 (2003) [doi:10.1016/S0379-6779(02)00603-3].

[54]   T. A. Skotheim, *Handbook of Conducting Polymers*, M. Dekker (1998).

[55]   S. D. Baranovskii, M. Wiemer, A. V. Nenashev, F. Jansson, and F. Gebhard, "Calculating the Efficiency of Exciton Dissociation at the Interface between a Conjugated Polymer and an Electron Acceptor," *J. Phys. Chem. Lett.* **3**(9), 1214–1221 (2012) [doi:10.1021/jz300123k].

[56]   J.-L. Brédas, J. Cornil, and A. J. Heeger, "The exciton binding energy in luminescent conjugated polymers," *Advanced Materials* **8**(5), 447–452 (1996) [doi:10.1002/adma.19960080517].

[57]   O. G. Reid, J. A. N. Malik, G. Latini, S. Dayal, N. Kopidakis, C. Silva, N. Stingelin, and G. Rumbles, "The influence of solid-state microstructure on the origin and yield of long-lived photogenerated charge in neat semiconducting polymers," *Journal of Polymer Science Part B: Polymer Physics* **50**(1), 27–37 (2012) [doi:10.1002/polb.22379].

[58]   I. G. Scheblykin, A. Yartsev, T. Pullerits, V. Gulbinas, and V. Sundström, "Excited State and Charge Photogeneration Dynamics in Conjugated

Polymers," *J. Phys. Chem. B* **111**(23), 6303–6321 (2007) [doi:10.1021/jp068864f].

[59] D. E. Markov, E. Amsterdam, P. W. M. Blom, A. B. Sieval, and J. C. Hummelen, "Accurate Measurement of the Exciton Diffusion Length in a Conjugated Polymer Using a Heterostructure with a Side-Chain Cross-Linked Fullerene Layer," *J. Phys. Chem. A* **109**(24), 5266–5274 (2005) [doi:10.1021/jp0509663].

[60] D. W. McBranch, B. Kraabel, S. Xu, R. S. Kohlman, V. I. Klimov, D. D. C. Bradley, B. R. Hsieh, and M. Rubner, "Signatures of excitons and polaron pairs in the femtosecond excited-state absorption spectra of phenylene-based conjugated polymers and oligomers," *Synthetic Metals* **101**(1–3), 291–294 (1999) [doi:10.1016/S0379-6779(98)00373-7].

[61] B. Kraabel, D. McBranch, N. S. Sariciftci, D. Moses, and A. J. Heeger, "Ultrafast spectroscopic studies of photoinduced electron transfer from semiconducting polymers to C_{60}," *Phys. Rev. B* **50**(24), 18543 (1994) [doi:10.1103/PhysRevB.50.18543].

[62] R. Österbacka, M. Wohlgenannt, M. Shkunov, D. Chinn, and Z. V. Vardeny, "Excitons, polarons, and laser action in poly(p-phenylene vinylene) films," *The Journal of Chemical Physics* **118**(19), 8905–8916 (2003) [doi:doi:10.1063/1.1566937].

[63] S. R. Scully and M. D. McGehee, "Effects of optical interference and energy transfer on exciton diffusion length measurements in organic semiconductors," *Journal of Applied Physics* **100**(3), 034907–034907–5 (2006) [doi:10.1063/1.2226687].

[64] Z. Wang, S. Mazumdar, and A. Shukla, "Photophysics of charge-transfer excitons in thin films of π-conjugated polymers," *Phys. Rev. B* **78**(23), 235109 (2008) [doi:10.1103/PhysRevB.78.235109].

[65] S. V. Chasteen, J. O. Härter, G. Rumbles, J. C. Scott, Y. Nakazawa, M. Jones, H.-H. Hörhold, H. Tillman, and S. A. Carter, "Comparison of blended versus layered structures for poly(p-phenylene vinylene)-based polymer photovoltaics," *Journal of Applied Physics* **99**(3), 033709–033709–10 (2006) [doi:10.1063/1.2168046].

[66] T. Förster, "Zwischenmolekulare Energiewanderung und Fluoreszenz," *Annalen der Physik* **437**(1-2), 55–75 (1948) [doi:10.1002/andp.19484370105].

[67] D. L. Dexter, "A Theory of Sensitized Luminescence in Solids," *The Journal of Chemical Physics* **21**(5), 836–850 (1953) [doi:10.1063/1.1699044].

[68] D. Beljonne, G. Pourtois, C. Silva, E. Hennebicq, L. M. Herz, R. H. Friend, G. D. Scholes, S. Setayesh, K. Müllen, et al., "Interchain vs. intrachain energy transfer in acceptor-capped conjugated polymers," *PNAS* **99**(17), 10982–10987 (2002) [doi:10.1073/pnas.172390999].

[69] L. Lüer, H.-J. Egelhaaf, D. Oelkrug, G. Cerullo, G. Lanzani, B.-H. Huisman, and D. de Leeuw, "Oxygen-induced quenching of photoexcited states in polythiophene films," *Organic Electronics* **5**(1–3), 83–89 (2004) [doi:10.1016/j.orgel.2003.12.005].

[70] J. E. Kroeze, T. J. Savenije, M. J. W. Vermeulen, and J. M. Warman, "Contactless Determination of the Photoconductivity Action Spectrum, Exciton Diffusion Length, and Charge Separation Efficiency in Polythiophene-Sensitized TiO2 Bilayers," *J. Phys. Chem. B* **107**(31), 7696–7705 (2003) [doi:10.1021/jp0217738].

[71] L. Onsager, "Initial Recombination of Ions," *Phys. Rev.* **54**(8), 554–557 (1938) [doi:10.1103/PhysRev.54.554].

[72] E. A. Silinsh and H. Inokuchi, "On charge carrier photogeneration mechanisms in organic molecular crystals," *Chemical Physics* **149**(3), 373–383 (1991) [doi:10.1016/0301-0104(91)90037-T].

[73] J. G. Müller, U. Lemmer, J. Feldmann, and U. Scherf, "Precursor States for Charge Carrier Generation in Conjugated Polymers Probed by Ultrafast Spectroscopy," *Phys. Rev. Lett.* **88**(14), 147401 (2002) [doi:10.1103/PhysRevLett.88.147401].

[74] T. Strobel, C. Deibel, and V. Dyakonov, "Role of Polaron Pair Diffusion and Surface Losses in Organic Semiconductor Devices," *Phys. Rev. Lett.* **105**(26), 266602 (2010) [doi:10.1103/PhysRevLett.105.266602].

[75] Y. Yi, V. Coropceanu, and J.-L. Brédas, "Exciton-Dissociation and Charge-Recombination Processes in Pentacene/C60 Solar Cells: Theoretical Insight into the Impact of Interface Geometry," *J. Am. Chem. Soc.* **131**(43), 15777–15783 (2009) [doi:10.1021/ja905975w].

[76] M. Tachiya, "Breakdown of the Onsager theory of geminate ion recombination," *The Journal of Chemical Physics* **89**(11), 6929–6935 (1988) [doi:10.1063/1.455317].

[77] S. Barth and H. Bässler, "Intrinsic Photoconduction in PPV-Type Conjugated Polymers," *Phys. Rev. Lett.* **79**(22), 4445–4448 (1997) [doi:10.1103/PhysRevLett.79.4445].

[78] V. I. Arkhipov, E. V. Emelianova, and H. Bässler, "Hot Exciton Dissociation in a Conjugated Polymer," *Phys. Rev. Lett.* **82**(6), 1321–1324 (1999) [doi:10.1103/PhysRevLett.82.1321].

[79]   V. I. Arkhipov, E. V. Emelianova, S. Barth, and H. Bässler, "Ultrafast on-chain dissociation of hot excitons in conjugated polymers," *Phys. Rev. B* **61**(12), 8207–8214 (2000) [doi:10.1103/PhysRevB.61.8207].

[80]   D. M. Basko and E. M. Conwell, "Hot exciton dissociation in conjugated polymers," *Phys. Rev. B* **66**(15), 155210 (2002) [doi:10.1103/PhysRevB.66.155210].

[81]   P. B. Miranda, D. Moses, and A. J. Heeger, "Ultrafast photogeneration of charged polarons in conjugated polymers," *Phys. Rev. B* **64**(8), 081201 (2001) [doi:10.1103/PhysRevB.64.081201].

[82]   P. B. Miranda, D. Moses, and A. J. Heeger, "Ultrafast charge photogeneration in conjugated polymers," *Synthetic Metals* **119**(1–3), 619–620 (2001) [doi:10.1016/S0379-6779(00)00817-1].

[83]   D. Moses, A. Dogariu, and A. J. Heeger, "Ultrafast detection of charged photocarriers in conjugated polymers," *Phys. Rev. B* **61**(14), 9373 (2000) [doi:10.1103/PhysRevB.61.9373].

[84]   R. Kersting, U. Lemmer, R. F. Mahrt, K. Leo, H. Kurz, H. Bässler, and E. O. Göbel, "Femtosecond energy relaxation in π-conjugated polymers," *Phys. Rev. Lett.* **70**(24), 3820–3823 (1993) [doi:10.1103/PhysRevLett.70.3820].

[85]   R. Kersting, U. Lemmer, M. Deussen, H. J. Bakker, R. F. Mahrt, H. Kurz, V. I. Arkhipov, H. Bässler, and E. O. Göbel, "Ultrafast Field-Induced Dissociation of Excitons in Conjugated Polymers," *Phys. Rev. Lett.* **73**(10), 1440–1443 (1994) [doi:10.1103/PhysRevLett.73.1440].

[86]   G. Cerullo, S. Stagira, M. Nisoli, S. De Silvestri, G. Lanzani, G. Kranzelbinder, W. Graupner, and G. Leising, "Excited-state dynamics of poly(para-phenylene)-type ladder polymers at high photoexcitation density," *Phys. Rev. B* **57**(20), 12806–12811 (1998) [doi:10.1103/PhysRevB.57.12806].

[87]   I. G. Austin and N. F. Mott, "Polarons in crystalline and non-crystalline materials," *Advances in Physics* **18**(71), 41–102 (1969) [doi:10.1080/00018736900101267].

[88]   J. L. Bredas and G. B. Street, "Polarons, bipolarons, and solitons in conducting polymers," *Acc. Chem. Res.* **18**(10), 309–315 (1985) [doi:10.1021/ar00118a005].

[89]   K. Fesser, A. R. Bishop, and D. K. Campbell, "Optical absorption from polarons in a model of polyacetylene," *Phys. Rev. B* **27**(8), 4804 (1983) [doi:10.1103/PhysRevB.27.4804].

[90] M. J. Rice and S. R. Phillpot, "Polarons and bipolarons in a model tetrahedrally bonded homopolymer," *Phys. Rev. Lett.* **58**(9), 937–940 (1987) [doi:10.1103/PhysRevLett.58.937].

[91] M. Deussen and H. Bässler, "Anion and cation absorption spectra of conjugated oligomers and polymers," *Chemical Physics* **164**(2), 247–257 (1992) [doi:10.1016/0301-0104(92)87148-3].

[92] J. Cornil, D. Beljonne, and J. L. Brédas, "Nature of optical transitions in conjugated oligomers. I. Theoretical characterization of neutral and doped oligo(phenylenevinylene)s," *The Journal of Chemical Physics* **103**(2), 834–841 (1995) [doi:10.1063/1.470116].

[93] J. Cornil, D. Beljonne, and J. L. Brédas, "Nature of optical transitions in conjugated oligomers. II. Theoretical characterization of neutral and doped oligothiophenes," *The Journal of Chemical Physics* **103**(2), 842–849 (1995) [doi:doi:10.1063/1.470065].

[94] S. Barth, H. Bässler, H. Rost, and H. H. Hörhold, "Extrinsic and intrinsic dc photoconductivity in a conjugated polymer," *Phys. Rev. B* **56**(7), 3844–3851 (1997) [doi:10.1103/PhysRevB.56.3844].

[95] P. A. Lane, X. Wei, and Z. V. Vardeny, "Spin and spectral signatures of polaron pairs in pi -conjugated polymers," *Phys. Rev. B* **56**(8), 4626 (1997) [doi:10.1103/PhysRevB.56.4626].

[96] P. J. Brown, H. Sirringhaus, M. Harrison, M. Shkunov, and R. H. Friend, "Optical spectroscopy of field-induced charge in self-organized high mobility poly(3-hexylthiophene)," *Phys. Rev. B* **63**(12), 125204 (2001) [doi:10.1103/PhysRevB.63.125204].

[97] R. Österbacka, X. M. Jiang, C. P. An, B. Horovitz, and Z. V. Vardeny, "Photoinduced Quantum Interference Antiresonances in π-Conjugated Polymers," *Phys. Rev. Lett.* **88**(22), 226401 (2002) [doi:10.1103/PhysRevLett.88.226401].

[98] X. M. Jiang, R. Österbacka, O. Korovyanko, C. P. An, B. Horovitz, R. A. . Janssen, and Z. V. Vardeny, "Spectroscopic Studies of Photoexcitations in Regioregular and Regiorandom Polythiophene Films," *Advanced Functional Materials* **12**(9), 587–597 (2002) [doi:10.1002/1616-3028(20020916)12:9<587::AID-ADFM587>3.0.CO;2-T].

[99] Beljonne, J. Cornil, H. Sirringhaus, P. J. Brown, M. Shkunov, R. H. Friend, and J.-L. Brødas, "Optical Signature of Delocalized Polarons in Conjugated Polymers," *Advanced Functional Materials* **11**(3), 229–234 (2001) [doi:10.1002/1616-3028(200106)11:3<229::AID-ADFM229>3.0.CO;2-L].

[100] J. M. Oberski, A. Greiner, and H. Bässler, "Absorption spectra of the anions of phenylenevinylene oligomers and polymer," *Chemical Physics Letters* **184**(5–6), 391–397 (1991) [doi:10.1016/0009-2614(91)80007-K].

[101] K. E. Ziemelis, A. T. Hussain, D. D. C. Bradley, R. H. Friend, J. Rühe, and G. Wegner, "Optical spectroscopy of field-induced charge in poly(3-hexyl thienylene) metal-insulator-semiconductor structures: Evidence for polarons," *Phys. Rev. Lett.* **66**(17), 2231–2234 (1991) [doi:10.1103/PhysRevLett.66.2231].

[102] G. f. Brown and J. Wu, "Third generation photovoltaics," *Laser & Photonics Reviews* **3**(4), 394–405 (2009) [doi:10.1002/lpor.200810039].

[103] P. Würfel, *Physics of Solar Cells: From Basic Principles to Advanced Concepts*, 2. aktualis. u. erg. Auflage, Wiley-VCH Verlag GmbH & Co. KGaA (2009).

[104] W. Shockley and H. J. Queisser, "Detailed Balance Limit of Efficiency of p-n Junction Solar Cells," *Journal of Applied Physics* **32**(3), 510–519 (1961) [doi:10.1063/1.1736034].

[105] "http://rredc.nrel.gov/solar/spectra/am1.5/."

[106] C. Winder and N. S. Sariciftci, "Low bandgap polymers for photon harvesting in bulk heterojunction solar cells," *J. Mater. Chem.* **14**(7), 1077 (2004) [doi:10.1039/b306630d].

[107] P. Würfel, "Thermodynamic limitations to solar energy conversion," *Physica E: Low-dimensional Systems and Nanostructures* **14**(1–2), 18–26 (2002) [doi:10.1016/S1386-9477(02)00355-7].

[108] R. T. Ross and A. J. Nozik, "Efficiency of hot-carrier solar energy converters," *Journal of Applied Physics* **53**(5), 3813–3818 (1982) [doi:10.1063/1.331124].

[109] P. T. Landsberg, H. Nussbaumer, and G. Willeke, "Band-band impact ionization and solar cell efficiency," *Journal of Applied Physics* **74**(2), 1451–1452 (1993) [doi:10.1063/1.354886].

[110] J. H. Werner, R. Brendel, and H.-J. Queisser, "Radiative efficiency limit of terrestrial solar cells with internal carrier multiplication," *Applied Physics Letters* **67**(7), 1028–1030 (1995) [doi:10.1063/1.114719].

[111] M. Grätzel, "Dye-sensitized solar cells," *Journal of Photochemistry and Photobiology C: Photochemistry Reviews* **4**(2), 145–153 (2003) [doi:10.1016/S1389-5567(03)00026-1].

[112] W. U. Huynh, J. J. Dittmer, and A. P. Alivisatos, "Hybrid Nanorod-Polymer Solar Cells," *Science* **295**(5564), 2425–2427 (2002) [doi:10.1126/science.1069156].

[113] A. P. Alivisatos, "Semiconductor Clusters, Nanocrystals, and Quantum Dots," *Science* **271**(5251), 933 –937 (1996) [doi:10.1126/science.271.5251.933].

[114] D. Gebeyehu, C. J. Brabec, F. Padinger, T. Fromherz, S. Spiekermann, N. Vlachopoulos, F. Kienberger, H. Schindler, and N. S. Sariciftci, "Solid state dye-sensitized TiO2 solar cells with poly(3-octylthiophene) as hole transport layer," *Synthetic Metals* **121**(1–3), 1549–1550 (2001) [doi:10.1016/S0379-6779(00)01239-X].

[115] D. L. Morel, E. L. Stogryn, A. K. Ghosh, T. Feng, P. E. Purwin, R. F. Shaw, C. Fishman, G. R. Bird, and A. P. Piechowski, "Organic photovoltaic cells. Correlations between cell performance and molecular structure," *J. Phys. Chem.* **88**(5), 923–933 (1984) [doi:10.1021/j150649a019].

[116] C. W. Tang, "Two-layer organic photovoltaic cell," *Applied Physics Letters* **48**(2), 183–185 (1986) [doi:10.1063/1.96937].

[117] P. Peumans, A. Yakimov, and S. R. Forrest, "Small molecular weight organic thin-film photodetectors and solar cells," *Journal of Applied Physics* **93**(7), 3693–3723 (2003) [doi:doi:10.1063/1.1534621].

[118] P. Peumans, S. Uchida, and S. R. Forrest, "Efficient bulk heterojunction photovoltaic cells using small-molecular-weight organic thin films," *Nature* **425**(6954), 158–162 (2003) [doi:10.1038/nature01949].

[119] J. J. M. Halls, C. A. Walsh, N. C. Greenham, E. A. Marseglia, R. H. Friend, S. C. Moratti, and A. B. Holmes, "Efficient photodiodes from interpenetrating polymer networks," *Nature* **376**(6540), 498–500 (1995) [doi:10.1038/376498a0].

[120] A. B. Tamayo, B. Walker, and T.-Q. Nguyen*, "A Low Band Gap, Solution Processable Oligothiophene with a Diketopyrrolopyrrole Core for Use in Organic Solar Cells," *J. Phys. Chem. C* **112**(30), 11545–11551 (2008) [doi:10.1021/jp8031572].

[121] B. Walker, C. Kim, and T.-Q. Nguyen, "Small Molecule Solution-Processed Bulk Heterojunction Solar Cells†," *Chem. Mater.* **23**(3), 470–482 (2011) [doi:10.1021/cm102189g].

[122] "Heliatek sets new world record efficiency of 10.7% for its organic tandem cell," *Heliatek*, <http://www.heliatek.com/newscenter/latest_news/heliatek-erzielt-mit-

107-effizienz-neuen-weltrekord-fur-seine-organische-
tandemzelle/?lang=en> (26 October 2012).

[123] S. H. Park, A. Roy, S. Beaupré, S. Cho, N. Coates, J. S. Moon, D. Moses, M. Leclerc, K. Lee, et al., "Bulk heterojunction solar cells with internal quantum efficiency approaching 100%," *Nature Photonics* **3**(5), 297–302 (2009) [doi:10.1038/nphoton.2009.69].

[124] J. C. Hummelen, B. W. Knight, F. LePeq, F. Wudl, J. Yao, and C. L. Wilkins, "Preparation and Characterization of Fulleroid and Methanofullerene Derivatives," *J. Org. Chem.* **60**(3), 532–538 (1995) [doi:10.1021/jo00108a012].

[125] H. W. Kroto, J. R. Heath, S. C. O'Brien, R. F. Curl, and R. E. Smalley, "C60: Buckminsterfullerene," , *Published online: 14 November 1985; | doi:10.1038/318162a0* **318**(6042), 162–163 (1985) [doi:10.1038/318162a0].

[126] K. Schulze, C. Uhrich, R. Schüppel, K. Leo, M. Pfeiffer, E. Brier, E. Reinold, and P. Bäuerle, "Efficient Vacuum-Deposited Organic Solar Cells Based on a New Low-Bandgap Oligothiophene and Fullerene C60," *Advanced Materials* **18**(21), 2872–2875 (2006) [doi:10.1002/adma.200600658].

[127] J. Xue, S. Uchida, B. P. Rand, and S. R. Forrest, "4.2% efficient organic photovoltaic cells with low series resistances," *Applied Physics Letters* **84**(16), 3013–3015 (2004) [doi:10.1063/1.1713036].

[128] G. Yu, J. Gao, J. C. Hummelen, F. Wudl, and A. J. Heeger, "Polymer Photovoltaic Cells: Enhanced Efficiencies via a Network of Internal Donor-Acceptor Heterojunctions," *Science* **270**(5243), 1789–1791 (1995) [doi:10.1126/science.270.5243.1789].

[129] B. Walker, A. B. Tamayo, X.-D. Dang, P. Zalar, J. H. Seo, A. Garcia, M. Tantiwiwat, and T.-Q. Nguyen, "Flexible Organic Solar Cells: Nanoscale Phase Separation and High Photovoltaic Efficiency in Solution-Processed, Small-Molecule Bulk Heterojunction Solar Cells (Adv. Funct. Mater. 19/2009)," *Advanced Functional Materials* **19**(19), n/a–n/a (2009) [doi:10.1002/adfm.200990086].

[130] K. Walzer, B. Maennig, M. Pfeiffer, and K. Leo, "Highly Efficient Organic Devices Based on Electrically Doped Transport Layers," *Chem. Rev.* **107**(4), 1233–1271 (2007) [doi:10.1021/cr050156n].

[131] R. A. Marsh, C. Groves, and N. C. Greenham, "A microscopic model for the behavior of nanostructured organic photovoltaic devices," *Journal of*

Applied Physics **101**(8), 083509–083509–7 (2007) [doi:10.1063/1.2718865].

[132] J. Weickert, R. B. Dunbar, H. C. Hesse, W. Wiedemann, and L. Schmidt-Mende, "Nanostructured Organic and Hybrid Solar Cells," *Advanced Materials* **23**(16), 1810–1828 (2011) [doi:10.1002/adma.201003991].

[133] H. Hesse, "Supramolecular assembly and nanoscale morphologies for organic photovoltaic devices," Ludwig-Maximilians Universität München (2012).

[134] S. Günes, H. Neugebauer, and N. S. Sariciftci, "Conjugated Polymer-Based Organic Solar Cells," *Chem. Rev.* **107**(4), 1324–1338 (2007) [doi:10.1021/cr050149z].

[135] I.-W. Hwang, C. Soci, D. Moses, Z. Zhu, D. Waller, R. Gaudiana, C. J. Brabec, and A. J. Heeger, "Ultrafast Electron Transfer and Decay Dynamics in a Small Band Gap Bulk Heterojunction Material," *Adv. Mater.* **19**(17), 2307–2312 (2007) [doi:10.1002/adma.200602437].

[136] C. J. Brabec, A. Cravino, D. Meissner, N. S. Sariciftci, T. Fromherz, M. T. Rispens, L. Sanchez, and J. C. Hummelen, "Origin of the Open Circuit Voltage of Plastic Solar Cells," *Advanced Functional Materials* **11**(5), 374–380 (2001) [doi:10.1002/1616-3028(200110)11:5<374::AID-ADFM374>3.0.CO;2-W].

[137] D. Veldman, S. C. J. Meskers, and R. A. J. Janssen, "The Energy of Charge-Transfer States in Electron Donor–Acceptor Blends: Insight into the Energy Losses in Organic Solar Cells," *Advanced Functional Materials* **19**(12), 1939–1948 (2009) [doi:10.1002/adfm.200900090].

[138] C. Deibel and V. Dyakonov, "Polymer–fullerene bulk heterojunction solar cells," *Reports on Progress in Physics* **73**(9), 096401 (2010) [doi:10.1088/0034-4885/73/9/096401].

[139] C. Deibel, T. Strobel, and V. Dyakonov, "Role of the Charge Transfer State in Organic Donor–Acceptor Solar Cells," *Advanced Materials* **22**(37), 4097–4111 (2010) [doi:10.1002/adma.201000376].

[140] I. D. Parker, "Carrier tunneling and device characteristics in polymer light-emitting diodes," *Journal of Applied Physics* **75**(3), 1656–1666 (1994) [doi:10.1063/1.356350].

[141] B. A. Gregg and M. C. Hanna, "Comparing organic to inorganic photovoltaic cells: Theory, experiment, and simulation," *Journal of Applied Physics* **93**(6), 3605–3614 (2003) [doi:10.1063/1.1544413].

[142] S. E. Shaheen, C. J. Brabec, N. S. Sariciftci, F. Padinger, T. Fromherz, and J. C. Hummelen, "2.5% efficient organic plastic solar cells," *Applied Physics Letters* **78**(6), 841–843 (2001) [doi:10.1063/1.1345834].

[143] C. J. Brabec, N. S. Sariciftci, and J. C. Hummelen, "Plastic Solar Cells," *Advanced Functional Materials* **11**(1), 15–26 (2001) [doi:10.1002/1616-3028(200102)11:1<15::AID-ADFM15>3.0.CO;2-A].

[144] P. Schilinsky, C. Waldauf, and C. J. Brabec, "Recombination and loss analysis in polythiophene based bulk heterojunction photodetectors," *Applied Physics Letters* **81**(20), 3885–3887 (2002) [doi:10.1063/1.1521244].

[145] F. Padinger, R. s. Rittberger, and N. s. Sariciftci, "Effects of Postproduction Treatment on Plastic Solar Cells," *Advanced Functional Materials* **13**(1), 85–88 (2003) [doi:10.1002/adfm.200390011].

[146] M. Reyes-Reyes, K. Kim, and D. L. Carroll, "High-efficiency photovoltaic devices based on annealed poly(3-hexylthiophene) and 1-(3-methoxycarbonyl)-propyl-1- phenyl-(6,6)C61 blends," *Applied Physics Letters* **87**(8), 083506–083506-3 (2005) [doi:10.1063/1.2006986].

[147] K. G. Jespersen, W. J. D. Beenken, Y. Zaushitsyn, A. Yartsev, M. Andersson, T. Pullerits, and V. Sundström, "The electronic states of polyfluorene copolymers with alternating donor-acceptor units," *J. Chem. Phys.* **121**(24), 12613 (2004) [doi:10.1063/1.1817873].

[148] M. Svensson, F. Zhang, S. c. Veenstra, W. j. h. Verhees, J. c. Hummelen, J. m. Kroon, O. Inganäs, and M. r. Andersson, "High-Performance Polymer Solar Cells of an Alternating Polyfluorene Copolymer and a Fullerene Derivative," *Advanced Materials* **15**(12), 988–991 (2003) [doi:10.1002/adma.200304150].

[149] F. Zhang, K. G. Jespersen, C. Björström, M. Svensson, M. R. Andersson, V. Sundström, K. Magnusson, E. Moons, A. Yartsev, et al., "Influence of Solvent Mixing on the Morphology and Performance of Solar Cells Based on Polyfluorene Copolymer/Fullerene Blends," *Advanced Functional Materials* **16**(5), 667–674 (2006) [doi:10.1002/adfm.200500339].

[150] M.-H. Chen, J. Hou, Z. Hong, G. Yang, S. Sista, L.-M. Chen, and Y. Yang, "Efficient Polymer Solar Cells with Thin Active Layers Based on Alternating Polyfluorene Copolymer/Fullerene Bulk Heterojunctions," *Advanced Materials* **21**(42), 4238–4242 (2009) [doi:10.1002/adma.200900510].

[151] N. Blouin, A. Michaud, D. Gendron, S. Wakim, E. Blair, R. Neagu-Plesu, M. Belletête, G. Durocher, Y. Tao, et al., "Toward a Rational Design of

Poly(2,7-Carbazole) Derivatives for Solar Cells," *J. Am. Chem. Soc.* **130**(2), 732–742 (2008) [doi:10.1021/ja0771989].

[152] Z. Zhu, D. Waller, R. Gaudiana, M. Morana, D. Mühlbacher, M. Scharber, and C. Brabec, "Panchromatic Conjugated Polymers Containing Alternating Donor/Acceptor Units for Photovoltaic Applications," *Macromolecules* **40**(6), 1981–1986 (2007) [doi:10.1021/ma0623760].

[153] P. Coppo and M. L. Turner, "Cyclopentadithiophene based electroactive materials," *J. Mater. Chem.* **15**(11), 1123–1133 (2005) [doi:10.1039/B412143K].

[154] M. Zhang, H. N. Tsao, W. Pisula, C. Yang, A. K. Mishra, and K. Müllen, "Field-Effect Transistors Based on a Benzothiadiazole–Cyclopentadithiophene Copolymer," *J. Am. Chem. Soc.* **129**(12), 3472–3473 (2007) [doi:10.1021/ja0683537].

[155] Y. Sun, G. C. Welch, W. L. Leong, C. J. Takacs, G. C. Bazan, and A. J. Heeger, "Solution-processed small-molecule solar cells with 6.7% efficiency," *Nature Materials* **11**(1), 44–48 (2012) [doi:10.1038/nmat3160].

[156] A. H. Zewail, "Femtochemistry: Atomic-Scale Dynamics of the Chemical Bond Using Ultrafast Lasers (Nobel Lecture)," *Angewandte Chemie International Edition* **39**(15), 2586–2631 (2000) [doi:10.1002/1521-3773(20000804)39:15<2586::AID-ANIE2586>3.0.CO;2-O].

[157] G. Cerullo and S. De Silvestri, "Ultrafast optical parametric amplifiers," *Review of Scientific Instruments* **74**(1), 1 –18 (2003) [doi:10.1063/1.1523642].

[158] G. Cerullo, C. Manzoni, L. Lüer, and D. Polli, "Time-resolved methods in biophysics. 4. Broadband pump–probe spectroscopy system with sub-20 fs temporal resolution for the study of energy transfer processes in photosynthesis," *Photochem. Photobiol. Sci.* **6**(2), 135–144 (2007) [doi:10.1039/B606949E].

[159] R. Trebino, *Frequency-Resolved Optical Gating: The Measurement of Ultrashort Laser Pulses*, 1st ed., Springer (2002).

[160] M. Bradler, C. Homann, and E. Riedle, "Mid-IR femtosecond pulse generation on the microjoule level up to 5??m at high repetition rates," *Opt. Lett.* **36**(21), 4212–4214 (2011) [doi:10.1364/OL.36.004212].

[161] N. Demirdöven, M. Khalil, O. Golonzka, and A. Tokmakoff, "Dispersion compensation with optical materials for compression of intense sub-100-

fs mid-infrared pulses," *Opt. Lett.* **27**(6), 433–435 (2002) [doi:10.1364/OL.27.000433].

[162] R. Tautz, "Single-shot Characterization of Sub-10-fs Laser Pulses," Diploma Thesis, Universität Regensburg (2008).

[163] U. Megerle, I. Pugliesi, C. Schriever, C. F. Sailer, and E. Riedle, "Sub-50 fs broadband absorption spectroscopy with tunable excitation: putting the analysis of ultrafast molecular dynamics on solid ground," *Appl. Phys. B* **96**(2-3), 215–231 (2009) [doi:10.1007/s00340-009-3610-0].

[164] M. Bradler, P. Baum, and E. Riedle, "Femtosecond continuum generation in bulk laser host materials with sub-µJ pump pulses," *Appl. Phys. B* **97**(3), 561–574 (2009) [doi:10.1007/s00340-009-3699-1].

[165] D. Polli, L. Lüer, and G. Cerullo, "High-time-resolution pump-probe system with broadband detection for the study of time-domain vibrational dynamics," *Review of Scientific Instruments* **78**(10), 103108–103108–9 (2007) [doi:10.1063/1.2800778].

[166] R. Tautz, E. Da Como, C. Wiebeler, G. Soavi, I. Dumsch, N. Fröhlich, G. Grancini, S. Allard, U. Scherf, et al., "Charge Photogeneration in Donor-Acceptor Conjugated Materials: Influence of Excess Excitation Energy and Chain Length," *J. Am. Chem. Soc., in press* (2013) [DOI: 10.1021/ja309252a].

[167] Y. H. Kim, D. Spiegel, S. Hotta, and A. J. Heeger, "Photoexcitation and doping studies of poly(3-hexylthienylene)," *Phys. Rev. B* **38**(8), 5490–5495 (1988) [doi:10.1103/PhysRevB.38.5490].

[168] K. Suganandam, P. Santhosh, M. Sankarasubramanian, A. Gopalan, T. Vasudevan, and K.-P. Lee, "Fe3+ ion sensing characteristics of polydiphenylamine--electrochemical and spectroelectrochemical analysis," *Sensors and Actuators B: Chemical* **105**(2), 223–231 (2005) [doi:16/j.snb.2004.06.005].

[169] O. Khatib, B. Lee, J. Yuen, Z. Q. Li, M. Di Ventra, A. J. Heeger, V. Podzorov, and D. N. Basov, "Infrared signatures of high carrier densities induced in semiconducting poly(3-hexylthiophene) by fluorinated organosilane molecules," *J. Appl. Phys.* **107**(12), 123702 (2010) [doi:10.1063/1.3436567].

[170] I. C. Lewis and L. S. Singer, "Electron Spin Resonance of Radical Cations Produced by the Oxidation of Aromatic Hydrocarbons with SbCl5," *The Journal of Chemical Physics* **43**(8), 2712–2727 (1965) [doi:10.1063/1.1697200].

[171] L. Zaikowski, P. Kaur, C. Gelfond, E. Selvaggio, S. Asaoka, Q. Wu, H.-C. Chen, N. Takeda, A. R. Cook, et al., "Polarons, Bipolarons, and Side-By-Side Polarons in Reduction of Oligofluorenes," *J. Am. Chem. Soc.* **134**(26), 10852–10863 (2012) [doi:10.1021/ja301494n].

[172] I. Kriegel, C. Jiang, J. Rodríguez-Fernández, R. D. Schaller, D. V. Talapin, E. da Como, and J. Feldmann, "Tuning the Excitonic and Plasmonic Properties of Copper Chalcogenide Nanocrystals," *J. Am. Chem. Soc.* **134**(3), 1583–1590 (2012) [doi:10.1021/ja207798q].

[173] E. Itoh, K. Terashima, H. Nagai, and K. Miyairi, "Evaluation of poly(3-hexylthiophene)/polymeric insulator interface by charge modulation spectroscopy technique," *Thin Solid Films* **518**(2), 810–813 (2009) [doi:10.1016/j.tsf.2009.07.091].

[174] R. Steyrleuthner, S. Bange, and D. Neher, "Reliable electron-only devices and electron transport in n-type polymers," *Journal of Applied Physics* **105**, 064509 (2009) [doi:10.1063/1.3086307].

[175] C. K. Chan, F. Amy, Q. Zhang, S. Barlow, S. Marder, and A. Kahn, "N-type doping of an electron-transport material by controlled gas-phase incorporation of cobaltocene," *Chemical Physics Letters* **431**(1–3), 67–71 (2006) [doi:10.1016/j.cplett.2006.09.034].

[176] Y. Zhang, B. de Boer, and P. W. M. Blom, "Trap-free electron transport in poly(p-phenylene vinylene) by deactivation of traps with n-type doping," *Phys. Rev. B* **81**(8), 085201 (2010) [doi:10.1103/PhysRevB.81.085201].

[177] M. J. Frisch, G. W. Trucks, G. E. Scuseria, Robb, M. A., Cheeseman, J. R., Scalmani, G., Barone, V., Mennucci, B., Petersson, G. A., et al., *Gaussian 09, Revision B.01*, Gaussian, Inc., Wallingford CT (2009).

[178] Dennigtnon, R., Keith, T., and Millam, J., *GaussView*, Semichem Inc., Shawnee Mission KS (2009).

[179] D. Fazzi, G. Grancini, M. Maiuri, D. Brida, G. Cerullo, and G. Lanzani, "Ultrafast internal conversion in a low band gap polymer for photovoltaics: experimental and theoretical study," *Phys. Chem. Chem. Phys.* **14**(18), 6367–6374 (2012) [doi:10.1039/C2CP23917E].

[180] S. Fratiloiu, F. C. Grozema, and L. D. A. Siebbeles, "Optical Properties and Delocalization of Excess Negative Charges on Oligo(Phenylenevinylene)s: A Quantum Chemical Study," *The Journal of Physical Chemistry B* **109**(12), 5644–5652 (2005) [doi:10.1021/jp0443310].

[181] M. J. S. Dewar, E. G. Zoebisch, E. F. Healy, and J. J. P. Stewart, "Development and use of quantum mechanical molecular models. 76. AM1: a new general purpose quantum mechanical molecular model," *J. Am. Chem. Soc.* **107**(13), 3902–3909 (1985) [doi:10.1021/ja00299a024].

[182] N. Krebs, R. A. Probst, and E. Riedle, "Sub-20 fs pulses shaped directly in the UV by an acousto-optic programmable dispersive filter," *Opt. Express* **18**(6), 6164–6171 (2010) [doi:10.1364/OE.18.006164].

[183] C. Homann, P. Lang, and E. Riedle, "Generation of 30 fs pulses tunable from 189 to 240 nm with an all-solid-state setup," *J. Opt. Soc. Am. B* **29**(10), 2765–2769 (2012) [doi:10.1364/JOSAB.29.002765].

[184] C. Manzoni, D. Polli, and G. Cerullo, "Two-color pump-probe system broadly tunable over the visible and the near infrared with sub-30fs temporal resolution," *Review of Scientific Instruments* **77**(2), 023103 – 023103–023109 (2006) [doi:10.1063/1.2167128].

[185] D. Brida, M. Marangoni, C. Manzoni, S. D. Silvestri, and G. Cerullo, "Two-optical-cycle pulses in the mid-infrared from an optical parametric amplifier," *Opt. Lett.* **33**(24), 2901–2903 (2008) [doi:10.1364/OL.33.002901].

[186] D. Brida, G. Cirmi, C. Manzoni, S. Bonora, P. Villoresi, S. De Silvestri, and G. Cerullo, "Sub-two-cycle light pulses at 1.6 ?m from an optical parametric amplifier," *Opt. Lett.* **33**(7), 741–743 (2008) [doi:10.1364/OL.33.000741].

[187] T. Drori, C.-X. Sheng, A. Ndobe, S. Singh, J. Holt, and Z. V. Vardeny, "Below-Gap Excitation of pi -Conjugated Polymer-Fullerene Blends: Implications for Bulk Organic Heterojunction Solar Cells," *Phys. Rev. Lett.* **101**(3), 037401 (2008) [doi:10.1103/PhysRevLett.101.037401].

[188] C.-X. Sheng, M. Tong, S. Singh, and Z. V. Vardeny, "Experimental determination of the charge/neutral branching ratio eta in the photoexcitation of pi -conjugated polymers by broadband ultrafast spectroscopy," *Phys. Rev. B* **75**(8), 085206 (2007) [doi:10.1103/PhysRevB.75.085206].

[189] E. Da Como, N. J. Borys, P. Strohriegl, M. J. Walter, and J. M. Lupton, "Formation of a Defect-Free π-Electron System in Single β-Phase Polyfluorene Chains," *Journal of the American Chemical Society* **133**(11), 3690–3692 (2011) [doi:10.1021/ja109342t].

[190] N. Banerji, S. Cowan, E. Vauthey, and A. J. Heeger, "Ultrafast Relaxation of the Poly(3-hexylthiophene) Emission Spectrum," *The*

Journal of Physical Chemistry C **115**(19), 9726–9739 (2011) [doi:10.1021/jp1119348].

[191] F. Etzold, I. A. Howard, R. Mauer, M. Meister, T.-D. Kim, K.-S. Lee, N. S. Baek, and F. Laquai, "Ultrafast Exciton Dissociation Followed by Nongeminate Charge Recombination in PCDTBT:PCBM Photovoltaic Blends," *J. Am. Chem. Soc.* **133**(24), 9469–9479 (2011) [doi:10.1021/ja201837e].

[192] A. Köhler, D. A. dos Santos, D. Beljonne, Z. Shuai, J.-L. Brédas, A. B. Holmes, A. Kraus, K. Müllen, and R. H. Friend, "Charge separation in localized and delocalized electronic states in polymeric semiconductors," *Nature* **392**(6679), 903–906 (1998) [doi:10.1038/31901].

[193] H. A. Mizes and E. M. Conwell, "Photoinduced charge transfer in poly(p-phenylene vinylene)," *Phys. Rev. B* **50**(15), 11243–11246 (1994) [doi:10.1103/PhysRevB.50.11243].

[194] Z. Chen, M. Bird, V. Lemaur, G. Radtke, J. Cornil, M. Heeney, I. McCulloch, and H. Sirringhaus, "Origin of the different transport properties of electron and hole polarons in an ambipolar polyselenophene-based conjugated polymer," *Phys. Rev. B* **84**(11), 115211 (2011) [doi:10.1103/PhysRevB.84.115211].

[195] R. L. Martin, "Natural transition orbitals," *The Journal of Chemical Physics* **118**(11), 4775–4777 (2003) [doi:10.1063/1.1558471].

[196] D. K. Campbell, A. R. Bishop, and K. Fesser, "Polarons in quasi-one-dimensional systems," *Phys. Rev. B* **26**(12), 6862–6874 (1982) [doi:10.1103/PhysRevB.26.6862].

[197] P.-L. T. Boudreault, A. Najari, and M. Leclerc, "Processable Low-Bandgap Polymers for Photovoltaic Applications†," *Chem. Mater.* **23**(3), 456–469 (2011) [doi:10.1021/cm1021855].

[198] B. Souharce, "Triphenylamine and Carbazole-based Hole Transoprting Materials and their Apllications in Prganic Field-Effect Tranistors," Bergische Universität Wuppertal (2008).

[199] M. Wohlgenannt and Z. Valy Vardeny, "Photophysics properties of blue-emitting polymers," *Synthetic Metals* **125**(1), 55–63 (2001) [doi:10.1016/S0379-6779(01)00511-2].

[200] J. Guo, H. Ohkita, H. Benten, and S. Ito, "Charge Generation and Recombination Dynamics in Poly(3-hexylthiophene)/Fullerene Blend Films with Different Regioregularities and Morphologies," *Journal of the American Chemical Society* **132**(17), 6154–6164 (2010) [doi:10.1021/ja100302p].

[201] C. G. Shuttle, B. O'Regan, A. M. Ballantyne, J. Nelson, D. D. C. Bradley, and J. R. Durrant, "Bimolecular recombination losses in polythiophene: Fullerene solar cells," *Phys. Rev. B* **78**(11), 113201 (2008) [doi:10.1103/PhysRevB.78.113201].

[202] I. A. Howard, R. Mauer, M. Meister, and F. Laquai, "Effect of Morphology on Ultrafast Free Carrier Generation in Polythiophene:Fullerene Organic Solar Cells," *Journal of the American Chemical Society* **132**(42), 14866–14876 (2010) [doi:10.1021/ja105260d].

[203] J. Peet, J. Y. Kim, N. E. Coates, W. L. Ma, D. Moses, A. J. Heeger, and G. C. Bazan, "Efficiency enhancement in low-bandgap polymer solar cells by processing with alkane dithiols," *Nat Mater* **6**(7), 497–500 (2007) [doi:10.1038/nmat1928].

[204] P. W. M. Blom, V. D. Mihailetchi, L. J. A. Koster, and D. E. Markov, "Device Physics of Polymer:Fullerene Bulk Heterojunction Solar Cells," *Advanced Materials* **19**(12), 1551–1566 (2007) [doi:10.1002/adma.200601093].

[205] J. Y. Kim, K. Lee, N. E. Coates, D. Moses, T.-Q. Nguyen, M. Dante, and A. J. Heeger, "Efficient Tandem Polymer Solar Cells Fabricated by All-Solution Processing," *Science* **317**(5835), 222–225 (2007) [doi:10.1126/science.1141711].

[206] C. J. Brabec, G. Zerza, G. Cerullo, S. De Silvestri, S. Luzzati, J. C. Hummelen, and S. Sariciftci, "Tracing photoinduced electron transfer process in conjugated polymer/fullerene bulk heterojunctions in real time," *Chemical Physics Letters* **340**(3-4), 232–236 (2001) [doi:10.1016/S0009-2614(01)00431-6].

[207] J. Piris, T. E. Dykstra, A. A. Bakulin, P. H. M. van Loosdrecht, W. Knulst, M. T. Trinh, J. M. Schins, and L. D. A. Siebbeles, "Photogeneration and Ultrafast Dynamics of Excitons and Charges in P3HT/PCBM Blends," *The Journal of Physical Chemistry C* **113**(32), 14500–14506 (2009) [doi:10.1021/jp904229q].

[208] I.-W. Hwang, D. Moses, and A. J. Heeger, "Photoinduced Carrier Generation in P3HT/PCBM Bulk Heterojunction Materials," *J. Phys. Chem. C* **112**(11), 4350–4354 (2008) [doi:10.1021/jp075565x].

[209] N. Banerji, S. Cowan, M. Leclerc, E. Vauthey, and A. J. Heeger, "Exciton Formation, Relaxation, and Decay in PCDTBT," *J. Am. Chem. Soc.* **132**(49), 17459–17470 (2010) [doi:10.1021/ja105290e].

[210] R. A. Marsh, J. M. Hodgkiss, and R. H. Friend, "Direct Measurement of Electric Field-Assisted Charge Separation in Polymer:Fullerene Photovoltaic Diodes," *Advanced Materials* **22**(33), 3672–3676 (2010) [doi:10.1002/adma.201001010].

[211] S. De, T. Pascher, M. Maiti, K. G. Jespersen, T. Kesti, F. Zhang, O. Inganäs, A. Yartsev, and V. Sundström, "Geminate Charge Recombination in Alternating Polyfluorene Copolymer/Fullerene Blends," *J. Am. Chem. Soc.* **129**(27), 8466–8472 (2007) [doi:10.1021/ja068909q].

[212] B. Carsten, J. M. Szarko, H. J. Son, W. Wang, L. Lu, F. He, B. S. Rolczynski, S. J. Lou, L. X. Chen, et al., "Examining the Effect of the Dipole Moment on Charge Separation in Donor–Acceptor Polymers for Organic Photovoltaic Applications," *J. Am. Chem. Soc.* **133**(50), 20468–20475 (2011) [doi:10.1021/ja208642b].

[213] M. D. Irwin, D. B. Buchholz, A. W. Hains, R. P. H. Chang, and T. J. Marks, "p-Type semiconducting nickel oxide as an efficiency-enhancing anode interfacial layer in polymer bulk-heterojunction solar cells," *PNAS* **105**(8), 2783–2787 (2008) [doi:10.1073/pnas.0711990105].

[214] C. Soci, I.-W. Hwang, D. Moses, Z. Zhu, D. Waller, R. Gaudiana, C. J. Brabec, and A. J. Heeger, "Photoconductivity of a Low-Bandgap Conjugated Polymer," *Adv. Funct. Mater.* **17**(4), 632–636 (2007) [doi:10.1002/adfm.200600199].

[215] J. Guo, H. Ohkita, H. Benten, and S. Ito, "Near-IR Femtosecond Transient Absorption Spectroscopy of Ultrafast Polaron and Triplet Exciton Formation in Polythiophene Films with Different Regioregularities," *Journal of the American Chemical Society* **131**(46), 16869–16880 (2009) [doi:10.1021/ja906621a].

[216] R. Österbacka, C. P. An, X. M. Jiang, and Z. V. Vardeny, "Two-Dimensional Electronic Excitations in Self-Assembled Conjugated Polymer Nanocrystals," *Science* **287**(5454), 839–842 (2000) [doi:10.1126/science.287.5454.839].

[217] F. Paquin, G. Latini, M. Sakowicz, P.-L. Karsenti, L. Wang, D. Beljonne, N. Stingelin, and C. Silva, "Charge Separation in Semicrystalline Polymeric Semiconductors by Photoexcitation: Is the Mechanism Intrinsic or Extrinsic?," *Phys. Rev. Lett.* **106**(19), 197401 (2011) [doi:10.1103/PhysRevLett.106.197401].

[218] G. Dicker, M. P. de Haas, L. D. A. Siebbeles, and J. M. Warman, "Electrodeless time-resolved microwave conductivity study of charge-carrier photogeneration in regioregular poly(3-hexylthiophene) thin

films," *Phys. Rev. B* **70**(4), 045203 (2004) [doi:10.1103/PhysRevB.70.045203].

[219] L. Pandey, C. Risko, J. E. Norton, and J.-L. Brédas, "Donor–Acceptor Copolymers of Relevance for Organic Photovoltaics: A Theoretical Investigation of the Impact of Chemical Structure Modifications on the Electronic and Optical Properties," *Macromolecules* **45**(16), 6405–6414 (2012) [doi:10.1021/ma301164e].

[220] E. Collini and G. D. Scholes, "Coherent Intrachain Energy Migration in a Conjugated Polymer at Room Temperature," *Science* **323**(5912), 369–373 (2009) [doi:10.1126/science.1164016].

[221] J. Cornil, I. Gueli, A. Dkhissi, J. C. Sancho-Garcia, E. Hennebicq, J. P. Calbert, V. Lemaur, D. Beljonne, and J. L. Brédas, "Electronic and optical properties of polyfluorene and fluorene-based copolymers: A quantum-chemical characterization," *J. Chem. Phys.* **118**(14), 6615 (2003) [doi:10.1063/1.1561054].

[222] D. Hertel and H. Bässler, "Photoconduction in amorphous organic solids," *Chemphyschem* **9**(5), 666–688 (2008) [doi:10.1002/cphc.200700575].

[223] C. Deibel, T. Strobel, and V. Dyakonov, "Origin of the Efficient Polaron-Pair Dissociation in Polymer-Fullerene Blends," *Phys. Rev. Lett.* **103**(3), 036402 (2009) [doi:10.1103/PhysRevLett.103.036402].

[224] K. M. Coakley and M. D. McGehee, "Conjugated Polymer Photovoltaic Cells," *Chem. Mater.* **16**(23), 4533–4542 (2004) [doi:10.1021/cm049654n].

[225] C. Arndt, U. Zhokhavets, M. Mohr, G. Gobsch, M. Al-Ibrahim, and S. Sensfuss, "Determination of polaron lifetime and mobility in a polymer/fullerene solar cell by means of photoinduced absorption," *Synthetic Metals* **147**(1–3), 257–260 (2004) [doi:10.1016/j.synthmet.2004.06.037].

[226] P. W. M. Blom, M. J. M. de Jong, and M. G. van Munster, "Electric-field and temperature dependence of the hole mobility in poly(p-phenylene vinylene)," *Phys. Rev. B* **55**(2), R656–R659 (1997) [doi:10.1103/PhysRevB.55.R656].

[227] D. Chirvase, Z. Chiguvare, M. Knipper, J. Parisi, V. Dyakonov, and J. C. Hummelen, "Electrical and optical design and characterisation of regioregular poly(3-hexylthiophene-2,5diyl)/fullerene-based heterojunction polymer solar cells," *Synthetic Metals* **138**(1–2), 299–304 (2003) [doi:10.1016/S0379-6779(03)00027-4].

[228] J.-F. Chang, H. Sirringhaus, M. Giles, M. Heeney, and I. McCulloch, "Relative importance of polaron activation and disorder on charge transport in high-mobility conjugated polymer field-effect transistors," *Phys. Rev. B* **76**(20), 205204 (2007) [doi:10.1103/PhysRevB.76.205204].

[229] M. C. Scharber, M. Koppe, J. Gao, F. Cordella, M. A. Loi, P. Denk, M. Morana, H.-J. Egelhaaf, K. Forberich, et al., "Influence of the Bridging Atom on the Performance of a Low-Bandgap Bulk Heterojunction Solar Cell," *Advanced Materials* **22**(3), 367–370 (2010) [doi:10.1002/adma.200900529].

[230] C. L. Braun, "Electric field assisted dissociation of charge transfer states as a mechanism of photocarrier production," *The Journal of Chemical Physics* **80**(9), 4157–4161 (1984) [doi:10.1063/1.447243].

[231] M. Wojcik and M. Tachiya, "Accuracies of the empirical theories of the escape probability based on Eigen model and Braun model compared with the exact extension of Onsager theory," *The Journal of Chemical Physics* **130**(10), 104107–104107–8 (2009) [doi:10.1063/1.3082005].

[232] A. Devižis, A. Serbenta, K. Meerholz, D. Hertel, and V. Gulbinas, "Ultrafast Dynamics of Carrier Mobility in a Conjugated Polymer Probed at Molecular and Microscopic Length Scales," *Phys. Rev. Lett.* **103**(2), 027404 (2009) [doi:10.1103/PhysRevLett.103.027404].

[233] A. A. Bakulin, D. Y. Parashchuk, P. van Loosdrecht, and M. S. Pshenichnikov, "Ultrafast polarisation spectroscopy of photoinduced charges in a conjugated polymer," *Quantum Electronics* **39**(7), 643–648 (2009) [doi:10.1070/QE2009v039n07ABEH014130].

[234] P. M. Zimmerman, Z. Zhang, and C. B. Musgrave, "Singlet fission in pentacene through multi-exciton quantum states," *Nature Chemistry* **2**(8), 648–652 (2010) [doi:10.1038/nchem.694].

[235] L. D. A. Siebbeles, "Organic solar cells: Two electrons from one photon," *Nature Chemistry* **2**(8), 608–609 (2010) [doi:10.1038/nchem.720].

[236] M. W. B. Wilson, A. Rao, J. Clark, R. S. S. Kumar, D. Brida, G. Cerullo, and R. H. Friend, "Ultrafast Dynamics of Exciton Fission in Polycrystalline Pentacene," *J. Am. Chem. Soc.* **133**(31), 11830–11833 (2011) [doi:10.1021/ja201688h].

[237] E. C. Greyson, B. R. Stepp, X. Chen, A. F. Schwerin, I. Paci, M. B. Smith, A. Akdag, J. C. Johnson, A. J. Nozik, et al., "Singlet Exciton Fission for Solar Cell Applications: Energy Aspects of Interchromophore Coupling†," *The Journal of Physical Chemistry B* **114**(45), 14223–14232 (2010) [doi:10.1021/jp909002d].

[238] N. M. Gabor, Z. Zhong, K. Bosnick, J. Park, and P. L. McEuen, "Extremely Efficient Multiple Electron-Hole Pair Generation in Carbon Nanotube Photodiodes," *Science* **325**(5946), 1367–1371 (2009) [doi:10.1126/science.1176112].

[239] S. Bange, U. Scherf, and J. M. Lupton, "Absence of Singlet Fission and Carrier Multiplication in a Model Conjugated Polymer: Tracking the Triplet Population through Phosphorescence," *J. Am. Chem. Soc.* **134**(4), 1946–1949 (2012) [doi:10.1021/ja2102289].

[240] P. Sreearunothai, A. C. Morteani, I. Avilov, J. Cornil, D. Beljonne, R. H. Friend, R. T. Phillips, C. Silva, and L. M. Herz, "Influence of Copolymer Interface Orientation on the Optical Emission of Polymeric Semiconductor Heterojunctions," *Phys. Rev. Lett.* **96**(11), 117403 (2006) [doi:10.1103/PhysRevLett.96.117403].